Teaching and Training for Global Engineering

IEEE Press
445 Hoes Lane
Piscataway, NJ 08854

IEEE Press Editorial Board
Tariq Samad, *Editor in Chief*

George W. Arnold	Ziaoou Li	Ray Perez
Giancarlo Fortino	Vladimir Lumelsky	Linda Shafer
Dmitry Goldgof	Pui-In Mak	Zidong Wang
Ekram Hossain	Jeffrey Nanzer	MengChu Zhou

Kenneth Moore, *Director of IEEE Book and Information Services (BIS)*

Technical Reviewer

Frank Lu, *University of Texas at Arlington*

Teaching and Training for Global Engineering

Perspectives on Culture and Professional Communication Practices

Edited by

Kirk St.Amant
Louisiana Tech University and University of Limerick

Madelyn Flammia
University of Central Florida

IEEE PCS Professional Engineering Communication Series

Copyright © 2016 by The Institute of Electrical and Electronics Engineers, Inc.

Published by John Wiley & Sons, Inc., Hoboken, New Jersey. All rights reserved.
Published simultaneously in Canada.

No part of this publication may be reproduced, stored in a retrieval system, or transmitted in any form or by any means, electronic, mechanical, photocopying, recording, scanning, or otherwise, except as permitted under Section 107 or 108 of the 1976 United States Copyright Act, without either the prior written permission of the Publisher, or authorization through payment of the appropriate per-copy fee to the Copyright Clearance Center, Inc., 222 Rosewood Drive, Danvers, MA 01923, (978) 750-8400, fax (978) 750-4470, or on the web at www.copyright.com. Requests to the Publisher for permission should be addressed to the Permissions Department, John Wiley & Sons, Inc., 111 River Street, Hoboken, NJ 07030, (201) 748-6011, fax (201) 748-6008, or online at http://www.wiley.com/go/permission.

Limit of Liability/Disclaimer of Warranty: While the publisher and author have used their best efforts in preparing this book, they make no representations or warranties with respect to the accuracy or completeness of the contents of this book and specifically disclaim any implied warranties of merchantability or fitness for a particular purpose. No warranty may be created or extended by sales representatives or written sales materials. The advice and strategies contained herein may not be suitable for your situation. You should consult with a professional where appropriate. Neither the publisher nor author shall be liable for any loss of profit or any other commercial damages, including but not limited to special, incidental, consequential, or other damages.

For general information on our other products and services or for technical support, please contact our Customer Care Department within the United States at (800) 762-2974, outside the United States at (317) 572-3993 or fax (317) 572-4002.

Wiley also publishes its books in a variety of electronic formats. Some content that appears in print may not be available in electronic formats. For more information about Wiley products, visit our web site at www.wiley.com.

Library of Congress Cataloging-in-Publication Data is available.

ISBN: 978-1-118-32802-6

Printed in the United States of America

10 9 8 7 6 5 4 3 2 1

To my mother, Joan Pelletier St.Amant, for instilling in me the value of education and the desire to teach, and to my daughters, Lily Catherine St.Amant and Isabelle Marie St.Amant, for being the inspiration for all I do.

– Kirk St.Amant

To my husband, Fred Klingenhagen, for his unfailing love and support, and to my father-in-law, Declan Klingenhagen, for his contribution to the engineering profession.

–Madelyn Flammia

Contents

A Note from the Series Editor, xvii
Foreword, xix
Acknowledgments, xxvii
Introduction, xxix

SECTION I   DESIGN CONTEXTS

1 The Imperative of Teaching Linguistics to Twenty-First-Century Professional Communicators   3
Bruce Maylath and Steven Hammer
- 1.1 Introduction, 4
- 1.2 Why Linguistics? The Dangers of Monolingualism, 5
- 1.3 Linguistic Knowledge—Where Did It Go?, 6
- 1.4 Linguistics for the Professional Engineer and Communicator, 8
  - 1.4.1 Linguistics for Translation, 9
  - 1.4.2 Who Is Ready; Who Is Not, 11
  - 1.4.3 Beyond the Classroom, 12
  - 1.4.4 Sociolinguistics and Pragmatics, 13
- 1.5 Conclusion, 15
- References, 16

2 Cultural Contexts in Document Design   19
Yiqin Wang and Dan Wang
- 2.1 The Challenges of International Communication, 19
- 2.2 Literature Review, 21
  - 2.2.1 Cultural Differences in Perception and Thought Patterns, 21
  - 2.2.2 Cultural Differences in Text Organization, 23
  - 2.2.3 Creating a Context for Understanding, 24
- 2.3 Study Design, 25
- 2.4 High- or Low-Context Culture and the Extent of Explicit Description, 26
  - 2.4.1 High- or Low-Context Culture, 26

2.4.2 A Study of Chinese and German Automobile Literature, 26
2.4.3 Findings Observed, 28
2.5 Thought Pattern and Content Organization, 29
    2.5.1 Thought Pattern, 29
    2.5.2 A Study of Chinese and German Automobile Literature, 30
    2.5.3 Findings Observed, 32
2.6 Cultural Contexts in Text–Graphic Relationships, 33
    2.6.1 Text–Graphic Relationships, 33
    2.6.2 Cultural Factors and Their Impacts on Text–Graphic Relationships, 33
    2.6.3 A Study of Chinese and German Automobile Literature, 34
    2.6.4 Findings Observed, 37
2.7 Cultural Backgrounds, 38
    2.7.1 Aspects of Language, 38
    2.7.2 Thought Patterns, 39
    2.7.3 High-Context versus Low-Context Communication, 39
2.8 Applying Ideas to Training in Technical and Professional Communication, 39
    2.8.1 Applying Ideas, 40
    2.8.2 An Overview of the Proposed Exercise, 40
2.9 Conclusion, 42
References, 43

3 Teaching Image Standards in a Post-Globalization Age   47
Audrey G. Bennett

3.1 Image Design and Consumption in a Post-Globalization Age, 48
    3.1.1 The Changing Landscape, 48
    3.1.2 Agents of Change, 48
3.2 Socially Conscious Communication Design and the Evolution of Image Standards, 49
    3.2.1 Approaches to Images as Tools for Social Action, 50
    3.2.2 Connections to Pedagogy, 50
    3.2.3 Expanding to the Global Context, 51
3.3 Standards for Communicatively Effective Images, 52
    3.3.1 A Communicatively Effective Image Establishes Credibility, 52
    3.3.2 Communicatively Effective Images and Use in Public Contexts, 52
    3.3.3 The Design Process of a Communicatively Effective Image Includes the User, 53
    3.3.4 A Communicatively Effective Image Resonates With the Culture(s) of Users, 55
    3.3.5 A Communicatively Effective Image Sustains Humanity or the Environment, 56

3.4 Implementing Objective Metrics in Technical and Professional Communication Classes, 56
    3.4.1 Conventional Approaches to Evaluating Image Design, 56
    3.4.2 A Revised Approach to Image Evaluation, 57
    3.4.3 Connecting Assessment Metrics to Teaching Practices, 62
3.5 Conclusion, 63
References, 64

SECTION II SOCIETAL CONTEXTS

4 Linux on the Education Desktop: Bringing the "Glocal" into the Technical Communication Classroom   69

Brian D. Ballentine

4.1 Introduction, 69
4.2 Linux—Dominance and Absence in Different Markets, 70
    4.2.1 Linux and Server Markets, 71
    4.2.2 Linux and Handheld Devices, 71
    4.2.3 Linux and Licensing, 72
    4.2.4 Linux and Open Source Community Practices, 73
4.3 Linux on the Desktop, 74
    4.3.1 The Global Market, 74
    4.3.2 Open Source in Educational Contexts, 75
    4.3.3 Competition for the Education Market, 77
4.4 Aggressive Evangelism, 78
    4.4.1 Approaches to Evangelism, 79
    4.4.2 Considering Contexts, 79
4.5 Extremadura, 80
    4.5.1 The Regional Decision, 80
    4.5.2 Examining the Context, 81
4.6 The Glocal, 82
4.7 Situating Professional Communication Students in the Glocal, 82
4.8 Using Linux on the Desktop, 83
4.9 Conclusion, 87
References, 88

5 Teaching the Ethics of Intercultural Communication   91

Dan Voss and Bethany Aguad

5.1 Introduction: Globalization Introduces an Intercultural Dimension to Business Ethics, 92
    5.1.1 Value Analysis Superimposes Cultural "Filters" on Classical Ethical and Contemporary Value Models, 92
    5.1.2 Teaching the Ethics of Intercultural Communication Applies to Technical Students and Professionals, 93

5.2 Literature Review Represents the Intersection of Ethics, Intercultural Communication, and Science/Engineering, 93
    5.2.1 Defining Ethics Becomes More Complex in an Intercultural Context, 93
    5.2.2 Teaching Ethics Demands a Systematic, Analytical Approach to a Subjective Topic, 95
    5.2.3 Teaching Ethics within the Science/Engineering Community Raises Sensitivity to the Complexity of the Subject, 96
    5.2.4 Teaching Ethics for the International Technical Community Requires Awareness of Variations Across Cultures, 96
    5.2.5 Working on International Teams Hones Engineering Students' Intercultural Communication Skills, 97
5.3 Four Classical Ethical Models Form the Foundation for Studying the Ethics of Intercultural Communication, 98
    5.3.1 Aristotle, Confucius, Kant, and Mill Offer Four Fundamental Ethical Constructs, 99
    5.3.2 The Path Between Universalism and Relativism Leads into a Thorny Ethical Thicket, 100
    5.3.3 Contextual Relativism Strikes a Balance Between the Extremes of Blind Absolutism and Situational Ethics, 101
    5.3.4 Value Analysis Provides a Systematic Approach to Identifying and Resolving Ethical Conflicts, 102
5.4 Two Value Models Help Rank Values to Resolve Conflicts in Favor of the Greatest Good or the Least Harm, 103
    5.4.1 In the Concentric-Ring Model, Core Values Are Paramount, 104
    5.4.2 The Hierarchical Model Highlights Intense Conflicts Between Legal, Societal, and Authority-Based Values, 105
5.5 Value Models within Technology-Based Companies and Professional Associations Offer Broad Ethical Perspectives, 106
5.6 Before Analyzing Ethical Conflicts in an Intercultural Context, Its Important to Understand the Cultural Differences Involved, 109
    5.6.1 Conflicts Between Authority-Based Values (Corporate, Political, National, and Religious) Can Be Explosive, 111
    5.6.2 Examining Models of Ethics in Cross-Cultural Contexts Often Pivots on the Precarious Balance of Relativism and Universalism, 113
5.7 Analyzing Case Histories via a Multi-Tiered Process of Ethical Models and Cultural Filters that Clarifies Ethical Conflicts, Defines Alternative Actions, and Predicts Outcomes, 116
    5.7.1 The Process Flow for Value Analysis is a Complex but Systematic Chart: Perfect for Engineers, 116

5.7.2 It's Tempting to Quantify the Ethics of Intercultural Communication into an Equation. All Suggestions Welcome!, 117
5.8 Suggestions for Integrating the Specialized Topic of this Chapter into Academic Courses and Industry Training Classes, 118
    5.8.1 Situation 1: SME Professor in Intercultural Communication Teaching Engineering or Science Students, 118
    5.8.2 Situation 2: SME Professor in Engineering or Science Teaching a Technical Content Course, 119
    5.8.3 Situation 3: Industry In-Service Education Class or Training Workshop in Intercultural Communication, 120
5.9 Conclusion: The Authors Invite Further Research and Contributions, 121
References, 121

SECTION III ONLINE CONTEXTS

6 Autonomous Learning and New Possibilities for Intercultural Communication in Online Higher Education in Mexico   127
César Correa Arias

6.1 Introduction, 128
6.2 The Nature and Characteristics of Autonomous Learning, 129
    6.2.1 Historical Development of Autonomous Learning, 130
    6.2.2 The Inclusion of a Paradigm of Constructivism in Pedagogical Methods, 130
    6.2.3 Autonomous Learning and New Approaches in Education, 132
    6.2.4 Curricular Experiences as a Result of a Constructivist Approach, 133
6.3 Understanding and Applying Autonomous Learning, 134
    6.3.1 Using Problem-Based Learning within Autonomous Learning, 134
    6.3.2 Connections to Curricular Experiences, 137
    6.3.3 Curricular Experiences and Educational Stories, 138
6.4 The Role of ICTs in Autonomous Learning, 139
6.5 The Culture of Autonomous Learning Inside Institutions of Higher Education, 140
    6.5.1 The Culture of Using ICTs to Build and Develop Curricular Experiences, 140
    6.5.2 Process of Evaluation within Educational Trajectories, 141
    6.5.3 Autonomous Learning, ICTs, and Knowledge Construction, 142

  6.5.4 Rethinking Evaluative Contexts and Approaches, 144
6.6 Conclusion, 145
References, 147

7 E-Learning and Technical Communication for International Audiences   149

Darina M. Slattery and Yvonne Cleary

7.1 Teaching Technical Communication and E-Learning: An Introduction, 149
7.2 An Overview of Learning Pedagogies, 150
  7.2.1 The Behaviorist Approach, 151
  7.2.2 The Cognitivist Approach, 152
  7.2.3 The Constructivist Approach, 154
7.3 Intercultural Communication Pedagogies, 155
  7.3.1 Collaboration, 156
  7.3.2 Information Design, 156
  7.3.3 Rhetorical Awareness, 157
7.4 The Irish Context for Technical Communication and E-Learning, 158
  7.4.1 Economic Context, 158
  7.4.2 Technical Communication, E-Learning and Localization within the Irish IT Industry, 159
7.5 The Configuration of our Program, 160
  7.5.1 The Structure of our MA Program, 161
  7.5.2 The Courses in our MA Program, 162
7.6 The Assignments in the MA Program, 164
  7.6.1 Virtual Team Assignments, 164
  7.6.2 Design and Development Assignments, 165
  7.6.3 Writing Assignments, 166
7.7 Connecting Student Work to Different Contexts, 167
  7.7.1 Rewriting a Poorly Written Text, 167
  7.7.2 Critiquing and Redesigning an Instruction Manual, 167
  7.7.3 Writing a Reflective Learning Blog, 168
7.8 Conclusion, 169
References, 169

8 Teaching and Training with a Flexible Module for Global Virtual Teams   173

Pam Estes Brewer

8.1 Introduction, 173
8.2 The Origins of the Approach Presented in This Chapter, 174
8.3 International Virtual Communication and Experiential Learning, 175
  8.3.1 Experiential Education Theory, 176
  8.3.2 Experiential Learning as it Complements Virtual Team Learning, 178

8.4 Teaching the Topic, 179
    8.4.1 The Module-Based Approach, 180
    8.4.2 Case 1. Class, Introduction to Professional Writing. Spring 2010. Face-to-Face Class Format. One Virtual Team Project—A Report. Partnered with Applied Foreign Languages Students at Ching Yun University (CYU), Taiwan, 181
    8.4.3 Case 2. Class, Business Writing. Fall 2010. Face-to-Face Class Format. One Virtual Team Project—A Report. Partnered with a Business Administration Class at CYU, Taiwan, 182
    8.4.4 Case 3. Class, Business Writing. Spring 2011. Hybrid Class Format. Two Virtual Team Projects—A Proposal and an Analytical Report. Partnered with a Tourism Class at Yerevan State Linguistic University (YSLU), Armenia, and a Management Class at CYU, Taiwan, 183
    8.4.5 Case 4. Class, Business Writing. Fall 2011. Online Class Format. One Virtual Team Project—Analytical Report. Partnered with an Economics Class at CYU, Taiwan, 184
8.5 Observations/Reflections/Theory Development for All Classes, 185
8.6 Global Virtual Team Teaching Module, 190
    8.6.1 Instructor Development, 190
    8.6.2 Foundations for Implementation, 190
    8.6.3 Making Contacts and Sharing Experiences, 192
    8.6.4 Experiences Applying Ideas and Approaches, 195
8.7 Conclusion, 195
References, 196

SECTION IV EDUCATIONAL CONTEXTS

9 Strategies for Developing International Professional Communication Products   201

Helen M. Grady

9.1 Introduction to International Technical Communication, 201
9.2 Review of the Literature, 202
    9.2.1 Course Texts, 202
    9.2.2 Growing Interest in the Field, 203
    9.2.3 Fundamental Processes and Practices, 203
    9.2.4 Connections to Professional Communication Practices, 204
9.3 The International Technical Communication Course, 204
    9.3.1 Course Focus, Goals, and Objectives, 205
    9.3.2 Course Content, 206

        9.3.3   Cultural Briefing Assignment, 207
        9.3.4   Feasibility Report Assignment, 210
  9.4   Conclusion, 215
  References, 216

10 Teaching Cultural Heuristics Through Narratives: A Transdisciplinary Approach   219
  Han Yu
  10.1 A Transdisciplinary Approach for Global Engineers, 219
  10.2 Overview of Cultural Heuristics, 220
        10.2.1 Edward T. Hall's Concepts of Time, Space, and Context, 220
        10.2.2 Geert Hofstede's Five-Dimension Heuristic, 221
        10.2.3 Fons Trompenaars and Charles Hampden-Turner's Seven-Dimension Heuristic, 222
  10.3 Critiques and Counter-Critiques of Cultural Heuristics: How to Move Forward from Misguided Debates, 222
        10.3.1 Misattributed and Unsystematic Critiques, 223
        10.3.2 Differences in Research Paradigms, 224
        10.3.3 Forward from the Critiques: A Transdisciplinary Approach to Integrate the Interpretive and the Functionalist Approaches, 226
  10.4 Overview of Cultural Narratives, 227
        10.4.1 Case Studies, 227
        10.4.2 Cross-Cultural Dialogs, 228
        10.4.3 Critical Incidents, 228
        10.4.4 Open Stories, 229
  10.5 Implement the Transdisciplinary Approach: Teach Cultural Heuristics Through Narratives, 230
        10.5.1 An Extended Example of a Case Study for Teaching Heuristics, 230
  10.6 Potential Limitations: How to Select Quality Cultural Narratives, 234
  10.7 Conclusion, 236
  References, 237

11 Assessing Intercultural Outcomes in Engineering Programs   239
  Darla K. Deardorff and Duane L. Deardorff
  11.1 Introduction, 240
  11.2 An Introduction to the Literature of Outcome Assessments, 241
        11.2.1 Thinking About Intercultural Assessment, 241
        11.2.2 Connections to Assessment, 242
  11.3 Exploring Some Limitations to Intercultural Assessment Research, 244

        11.3.1   Variation, 244
        11.3.2   Self-Orientation, 244
        11.3.3   Causality, 245
        11.3.4   Skills, 245
        11.3.5   Focus, 245
        11.3.6   Bias, 245
        11.3.7   Quality, 245
        11.3.8   Gaps, 246
11.4   Strategies for Quality Assessment of Intercultural Learning Outcomes, 246
        11.4.1   Establishing Definitions, 246
        11.4.2   Stating Goals and Measurable Outcomes, 247
        11.4.3   Writing Outcomes, 247
11.5   Developing an Assessment Plan, 249
        11.5.1   Defining Direct and Indirect Measures, 249
        11.5.2   Making Sense of Intercultural Assessment, 250
        11.5.3   Defining Intercultural Competence, 250
        11.5.4   Key Considerations, 251
        11.5.5   Frameworks and Models of Assessment, 252
11.6   Quality Assessment, 252
        11.6.1   Defining Intercultural/Global Competence, 252
        11.6.2   Developing a Comprehensive Assessment Plan, 253
        11.6.3   Involving a Team in Assessment Efforts, 253
        11.6.4   Ensuring Intercultural Learning is Intentionally Addressed and Supported, 254
11.7   Developing Intercultural Competence in Students, 254
        11.7.1   Considering the Curriculum, 254
        11.7.2   Considering Aspects of Non-formal Learning, 255
        11.7.3   Using Collected Assessment Data, 255
        11.7.4   Evaluating the Assessment Plan and Process, 256
11.8   An Example of Intercultural Assessment, 256
        11.8.1   Georgia Institute of Technology (Georgia Tech), 256
        11.8.2   Purdue University, 257
11.9   Assessing Intercultural Outcomes in Engineering Programs, 258
11.10 Conclusion, 258
References, 259

Biographies, 263
Index, 269

A Note from the Series Editor

The *Professional Engineering Communication* series grows by one more with this collection that was edited by Kirk St.Amant and Madelyn Flammia, titled *Teaching and Training for Global Engineering*. As part of this series, brought to you by the IEEE Professional Communication Society (PCS) and with Wiley-IEEE Press, we aim to bring to you a wealth of information, hard-earned knowledge, wise observations, and thoughtful perspectives about approaching engineering and technical communication teaching and training for global work.

Experienced practitioners will tell you that communication is the cornerstone to solid, technical, and successful international or global work. And, in order to get communication to function properly, it must be as planned for and thought about as the technical bits. This amazing collection of experts will bring perspectives and expertise that will enlighten your communication efforts at every level, at every part of the workflow.

Over the years, I have had the happy opportunity to use information from each and every one of the authors in this collection as I teach at universities and train professional engineers and technicians. In the past decade or so, I have also had the joy of meeting most of these authors face-to-face and learn from them in their talks and workshops. This book is a powerhouse of the best and brightest in technical and engineering communication perspectives today.

Take these viewpoints and nuggets of knowledge and transform them into something that will work for your specific situation. Use their expertise to your own advantage… as a place to begin to thoughtfully morph what you do into something even stronger than it already is.

This set of authors enriches greatly our series, which has a mandate to explore areas of communication practices and application as applied to the engineering, technical, and scientific professions. Including the realms of business, governmental agencies, academia, and other areas, this series has and will continue to develop perspectives about the state of communication issues and potential solutions when at all possible.

While theory has its place (in this book and this series), we always look to be a source where recommendations for action and activity can be found. All of the books in the fast-growing PEC series keep a steady eye on the applicable while acknowledging the contributions that analysis, research, and theory can provide to these efforts. There is a strong commitment from the Professional Communication Society of IEEE and Wiley to design a set of information and resources that can be carried directly into engineering firms, technology organizations, and academia alike.

For the series, we work with this philosophy: at the core of engineering, science, and technical work are problem solving and discovery. These tasks require, at all levels, talented and agile communication practices. We need to effectively gather, vet, analyze, synthesize, control, and design communication pieces in order for any meaningful work to get done. This book, with such a strong chorus of expert voices, contributes deeply to that vision for the series.

Traci Nathans-Kelly, Ph.D.
Series Editor

Foreword
One Person's Perspective on Culture and Communication Practices

Why Are You Reading This Book?

You're reading this book, very likely, because you are an academic or a trainer, a student, or a practitioner who understands the importance of the global reach and interconnectedness of business. It's a cliché of sorts to say that the world is a global village. However, it would be hard to imagine a job today or in the foreseeable future in which our ability to communicate well with coworkers, managers, and our customers all over the world is not critical to our success and to the success of the companies for whom we work.

Why Is It Important to Consider Culture?

Experience is the best teacher. My experience in being exposed to and confused by the complexities of communication across cultures taught me that I needed to get educated so that I could become a better interpreter of the communication contexts in which I found myself. In sharing some of my stories, I hope to demonstrate the importance of considering, and learning about, cultures different from our own.

My first story comes from my early days as a professor of business and technical writing at an engineering-focused college. Seeking the opportunity to gain real-world experience with engineers, I accepted a short-term consulting job to work with Japanese engineers at a manufacturing facility in Georgia. These engineers were in training for management positions in Japan, and the first step in their training was to serve on a 2-year assignment in the United States. The plant manager was from the United States, as were most of the workers at the plant.

The focus of the work was to help the Japanese engineers converse more comfortably and fluently with their coworkers and managers from the United States on both technical/engineering topics as well as any other topics that might be useful. The opportunity to work with Japanese engineers was exciting; the location of the plant, however, was challenging in that it required a two-hour drive each way. Nonetheless, I enjoyed getting to know the Japanese engineers and seeing them progress rapidly in their English proficiency. However, when the plant manager asked me whether I would be interested in extending my contract for a longer period of time, I declined because I had underestimated the time commitment and the need to focus on my responsibilities as a new professor.

At the celebratory dinner held at a local Japanese restaurant at the completion of this phase of their training, I told the Japanese engineers that I would not be continuing because of the long commute and my busy schedule. The next day I got a call from the US plant manager, sharing with me the deep disgrace the Japanese engineers felt and their desire to know what they might have done to offend me.

I was flabbergasted. I thought I had been very gentle in explaining the reason I was not renewing my arrangement with the client, and I thought I had been very encouraging about how well they were doing with their conversational English at this point. Clearly, I had missed some important sub-text, some aspect of communication that was hidden below the surface but nonetheless noted by the Japanese engineers. They had lost face with me, as well as the plant manager who had hired me, and I had been the cause of it.

Fast forward a decade to my next major exposure to a culture that was "foreign" to me, this time in China. Here are a few typical examples of puzzling exchanges I encountered, from among many that I could name.

> Context: Waiting in an airport in Western China for the late arrival of the one flight to Beijing for that day. Clear blue sky above. No plane in sight.
>
> Me: To the ticket agent: "Why is the plane delayed?"
>
> Ticket agent to me: Looking up at the sky... pause... then responding to me: "Maybe bad weather?"
>
> Context: On a tour bus in rural China.
>
> Me: To the tour guide: "What kind of vegetables are those in the field?"
>
> Tour guide to me: Looking out at the field and then back at me... pause... "Chinese vegetables."

In each of these exchanges, I knew something was happening that prompted these kinds of responses. I just didn't know what to call it or how to avoid such awkward situations in the future.

Later, after taking some seminars in Asian culture, I came to understood that it was, once again, face. When asked a question, the Chinese person with whom I was speaking wanted to answer with *something* rather than nothing. Not answering would disappoint the person who asked the question and embarrass the person who cannot answer, resulting in a loss of face on both sides of the communication.

> Through experiences such as these, I began to appreciate how the many and nuanced aspects of cultural communication expectations can affect most—if not all—of the exchanges that take place between individuals.

How Can Teaching Provide Learning Opportunities?

I found myself in China as a result of being selected by my university as the first professor in a teaching exchange with a Beijing-based Chinese engineering university of similar size and scope. In preparation for my semester-long stay in China, I studied Mandarin for six months, the result being that I was minimally conversant in Pinyin, the language system of modern China, which translates Chinese characters into Latin script. However, I could not read Chinese characters.

Although I had begun my cultural education by reading several travel guides and Chinese histories, for an understanding of the deeper, more complex meaning of communication exchanges, I was often in the dark. As it turned out, I was as much a student of culture as a teacher of English. Here are two stories from my teaching encounters that provided excellent learning opportunities for me and can be used as examples of the need to learn about other cultures as much as possible *before* immersion in them.

The first story occurred at the beginning of the semester, and it involved middle school English teachers who had been handpicked from all over China to attend a specialized year-long course in English. Selection for this class was considered to be prestigious because it was taught by a Western, English-speaking teacher (i.e., me for the first semester and a colleague from my university for the second semester).

I started the first class meeting by describing the classroom style generally associated with Western teaching and learning, and I suggested that, to help conceptualize this idea, we (the students and I) should try to interact in the Western style I had just described. So, for instance, I instructed the students to remain seated when they addressed me (rather than stand when speaking, which was the expected practice of students in the Chinese school system). I noted, as well, that they should not all speak at once when I asked a question (another cultural norm in the Chinese classroom). Rather, I explained, I would use the Western teaching style/approach of calling on each student, one at a time, and I expected them to respond individually.

All went well until I called on someone who did not know the answer to the question I asked. When that student rose to respond, there was a brief moment of silence and then everyone in the room chimed in to answer the question for her. After I repeatedly witnessed this group response to a question posed to an individual, I came to understand that the students' support of a single member was motivated by the desire to save the face of the individual and maintain harmony in the group. Their cultural mores were too deeply engrained to change at a mere suggestion from me.

The second story occurred at the end of the semester, at the point at which I had graded the students' final exams and it became clear that one student would not pass the course. After I distributed the results of the final exam to the students, I was approached by several of her classmates, the best students in the class, who pleaded her case in asking that the failing student be given a passing grade. It was at that point I learned that, in Chinese culture, everyone has face. If one member of a group loses face, all members of the group to which that person belongs also lose face. Because the students who came to me had status (face) from their high grades in the course, they took on the responsibility to intervene on behalf of their fellow student who did not have enough face/status to make such a request. As I was engrained in my own culture's focus on individual achievement earned through individual effort, I did not allow myself to be persuaded to change the student's grade so that she could pass the course. The result was a loss of face for all involved—the students *and* the teacher. I often wonder whether I made the wrong decision. In hindsight, I feel that I did. And, not surprisingly, the failing student returned with her classmates in the second semester of the program.

Another story comes from teaching the Chinese graduate students, who were taking a different English class of mine in preparation for the national exam they would have to pass at the end of the year. For this course, the school's administrators provided me with a prescribed textbook, and I was expected to use it systematically.

One day, I decided to diverge from the dry set of exercises at the end of the chapter and instead offered the students an assignment that would give them an opportunity to use some creativity to write an essay to demonstrate mastery of the grammar and vocabulary topics covered in the chapter. At the start of class, I asked the students to volunteer to read their essays to their peers. No one volunteered. So I followed the common US teaching practice of calling on someone to perform the requested task. The student I selected sheepishly informed me that he had not prepared the essay. I was shocked, as these were excellent, hardworking graduate students. I called on another student. Same response: she had not prepared the essay.

When I came to the realization that no one had prepared the essay, I was baffled. Later, I learned from one of the students who came to me privately that the reason he and his classmates had not prepared the essay was that they felt compelled to do the exercises at the back of the chapter *first*, as these were "required," and that left no time for the *extra* assignment I had given them.

My assignment for the students caused a loss of face all around. I was the teacher, and they wanted to comply with my instructions; but they also had to pass the national exam at the end of the year, which would be based on mastery of the exercises in the textbook. My failure to fully understand what the students' goal was had put me in the awkward position of adding to their already heavy burden of homework. Worse, it put them in the awkward position of appearing to be unprepared for class. As a result, we all lost face.

I never gave another "extra" assignment.

On the surface, these examples appear to be relatively harmless. But they suggest communication complexities that must be studied and learned. With this knowledge, we can effectively move from the simple effort of conversation to the complex negotiation of a business transaction.

How Can We Communicate Well in Global Teams and with Global Customers?

It is a fact of modern business that increasing numbers of us are working in globally distributed teams. With technology to support regular communication with teammates around the world, we can inadvertently find ourselves stepping blindly into problems of miscommunication through our voice in mobile or conference calls, our body language in video conferences, and our written communication in text messages, instant messaging, chat rooms, and emails.

> Moreover, superficial treatments of cross-cultural communication—often presented as a checklist of do's and do not's—cannot begin to scratch the surface of the need to understand deeply how culture shapes and affects experience.

In sum, not only do we need to learn how to be effective communicators in global teams, but we also need to learn how to design effective products for global markets, write support or instructional documentation that can be easily and effectively translated and localized, and create websites for international and culturally specific users. These are just a few of the ways in which technical and professional communicators will be called on to create products that not only do not offend but, more importantly, increase customer satisfaction and thus contribute to the global market share of employers.

How Can We Learn About the Effect of Culture on User Experience?

There are many ways to learn how culture affects experience, as the authors of these chapters present. Not to be overlooked, however, is the value of usability testing. When products have global users, usability testing can support the learning curve by helping expose some of the cultural differences that need to be addressed when products have a global market. Here's a story that shows the usefulness of testing in global contexts.

In an international study conducted remotely to users around the world, I had the experience of seeing cultural differences revealed in the way in which the participants presented themselves during the test sessions. They had been recruited for an hour-long session in which they were to set up an application and do some typical tasks with it.

Early in our first day of testing, my team and I were surprised to hear the manager (a test participant located in Singapore) talking to other people in the room with him. It soon became clear that these other people were assisting the manager in responding to the tasks we asked him to do. The observers from the international company, headquartered in the United States and the United Kingdom, were taken aback by this group participation, as they were interested only in the manager's experience, not in the support the manager might receive from others who might be assigned to carry out some of the tasks. The client's goal in this instance was to learn whether the managers could use the application in question without outside assistance. By assisting the manager with this task, the other participants were, in effect, nullifying the objective of the overall research project.

At this point in my cultural education, I was "schooled" in the cultural norms of collectivist cultures, of which this was one. As a result, I was not surprised at the manager's face-saving approach of using a "backup" to assist him in learning how to use the application. The client, however, was unhappy with this turn of events and wanted to intervene in order to ask the manager to work alone on completing the task—an option I did not allow. My goal was to help the client understand their users' experience, and this was clearly one of their users.

Later that same day, the clients and I observed the same approach taken by a manager and his support staff from Lisbon, Portugal. The third time it happened was with a US who was working for a US company, but who was originally from a different culture.

With these experiences, we began to see a pattern and to appreciate that the application had to address the needs of managers and their senior staff in collectivist cultures where relationships and cooperation are highly valued and highly supportive of positive face. In a later usability study with the same application, the same co-learning, face-saving approach was used, this time in the collectivist culture of Mexico. Thus, through this usability research, it became clear that the culture in which something is used affects how members of that culture will interact with or use it.

> Usability testing can help organizations understand how cultural factors affect product adoption, which will help these products meet with success in other cultures.

How Do We Know What We Don't Know?

From my experiences, I have learned that we need to learn as much as we can about other cultures and expose ourselves as much as possible to other cultures if we truly want to communicate effectively with counterparts and clients in the modern context of globalization. To achieve this objective, we must begin by realizing we should always have our cultural antenna up when we are in unfamiliar cultural situations. Education, be it self-generated or within the structure of a course or program, is central to learning and teaching about culture.

An important resource toward that end can be found in this edited collection. The chapters in this book can be used to broaden how we teach our students and train our clients, as well as how we prepare ourselves for effective cross-cultural communication. We can adopt or adapt the ideas and the exercises presented in this text in ways that can best suit our own individual purposes and contexts of use. We can also learn from the cultural communication strategies presented by the authors of this collection, and we can apply such ideas to a variety of cultural settings. The entries in this collection provide a map of sorts that readers can use to navigate on their journey to learn more about cultural factors that can influence professional communication practices around the globe.

One of my favorite expressions is from the Chinese philosopher Lao Tzu: A journey of a thousand miles begins with a single step. I find that I am on a journey that began a long time ago and hasn't ended yet. Curiosity and a thirst for knowledge and understanding keep me hungry to learn more, always more. I learned a great deal from the chapters in this book. I can put some of these strategies and exercises from this text to good use in my Global Strategies course in our M.S. degree in information design and communication. I can also use the ideas from this collection as I continue to work with and to learn from my clients, especially their users, in usability testing sessions.

I certainly do not have the key that unlocks cultural communication mysteries, but I can see through the keyhole because of a lifetime of learning. I'm sure you feel the same, or you would not be reading this excellent book.

Carol M. Barnum
Southern Polytechnic State University

Acknowledgments

This book would not have been possible without the help and support of a number of dedicated individuals who were essential to moving this project from an initial idea to a final, completed text.

To begin, we wish to thank Traci Nathans-Kelly, the *Professional Engineering Communication* Series Editor, for her continual guidance, editorial insights, and steadfast support for this project. Her seemingly endless time, patience, and work were essential to making this book possible. We also wish to thank Mary Hatcher, Editor at Wiley-IEEE Press, for her time and dedication to helping us navigate a range of publishing related items—from copyright to publication specifications. To this end, we also wish to thank the Professional Communication Society of the IEEE for creating a series such as this one that provides an important, new space in which ideas on engineering and communication can find a voice and begin a greater dialogue across fields.

Additionally, special thanks needs to be given to our contributing authors for their patience through the overall book publishing process—from the submission of initial manuscripts for review to the revision process to making final edits on chapters. We greatly appreciate your support for and your participation in this project.

Finally, we wish to thank our friends and families for their faith in this project and for providing the inspiration to conceive of the initial idea for this book and the motivation to see that idea realized.

Introduction
Rethinking the Context of Professional Communication Education for the Global Age

The Global Context of the Modern Workplace

Today's workplace is one where employees often operate in overlapping local and global contexts. In some cases, this overlap happens when individuals located in different nations collaborate on a common project. In other cases, this overlap occurs when local companies create products for or offer services to international markets. Moreover, advances in communication and collaboration technologies mean such local–global overlap can occur almost seamlessly—to the point at which individuals might not realize when they are working in local versus international contexts.

Success in this new workplace is often a matter of communication. Only through the effective exchange of information can the members of a team—be it locally based or globally distributed—move successfully toward a common objective. Only by gathering essential information on local and global audiences can organizations develop effective products and services for such audiences. The blurring between local and global, however, means it is increasingly difficult to determine when one is interacting in a local or an international context. In fact, given the interconnected nature of the modern economy, it is perhaps a moot point to think in terms of such distinctions.

To interact effectively in such environments, today's employees need to understand a range of linguistic, cultural, and technological factors that can affect professional interactions in different contexts. Doing so is no easy task. Yet, as the world becomes more interconnected, employees will increasingly need to hone their intercultural communication skills to meet the communication expectations of the blended local–global workplace.

The Educational Challenges Created by This Context

These new workplace expectations create an interesting situation for educators in professional communication. For many of these individuals, the area of intercultural communication is a new one that will require additional training to understand essential concepts and to teach related ideas. For others, the topic might not be new or unknown, but approaches to teaching it effectively are. Both kinds of situations reveal a need for resources that effectively address this topic within the context of different courses, curricula, and institutions.

One major issue to address becomes the scope of how to approach the overall topic—a classic case of the overlapping sentiments "Where do I begin?" "What should I cover?" and "How do I teach this topic?" This educational context reinforces the need for resources that can answer such questions and address the overlap among them.

> In sum, there is a need for a reference instructors can use to learn more about foundational topics in intercultural communication and approaches to teaching these topics effectively.

The objective of this text is to provide individuals with a resource they can use to begin addressing this need in a variety of educational contexts.

The Focus of the Collection within This Educational Context

No single text can effectively cover all topics, ideas, and approaches associated with teaching intercultural communication. (It would take a library full of texts to achieve this goal.) This collection should therefore be viewed as a starting point—an introductory text (vs. a comprehensive resource) designed to provide educators with a foundational understanding of how to teach international communication in different professional communication classes. Readers should, in turn, view this collection as an initial resource they can use to begin their own research on topics related to teaching intercultural factors associated with professional communication.

To create such an introductory resource, the editors have selected entries that cover many areas central to communicating effectively in international settings (and thus address the concerns of "Where do I begin?" and "What should I cover?"). These entries also provide a range of perspectives on, examples of, and concepts affecting the teaching of this topic within professional communication contexts. (In this way, they address the question of "How do I teach this topic?") Accordingly, the complexity of these topics means the treatment of certain factors varies from section to section within this collection. The overall combination of ideas, perspectives, and approaches covered in each section, however, provides readers with the foundation needed to integrate the teaching of intercultural concepts into professional communication classes and across an overall curriculum or program. With these ideas in mind, let's take a look at the organization of this collection.

The General Structure (Context) of This Collection

The chapters in this text have been organized into four context-based sections. Each section focuses on a specific theme related to international professional communication. The organization of the first three sections is designed to examine a specific factor that affects professional communication practices in international contexts. Through this structure, the chapters in each of these three sections introduce readers to central concepts associated with international communication. They also present approaches for teaching such concepts in professional communication classes. To do so, the chapters in these sections

- provide foundational information on a key topic (and address the concern of "Where do I begin?" and "What should I cover?")
- suggest strategies for teaching that specific topic (addressing the concern of "How do I teach this topic?")

This approach enhances the reader's knowledge of a particular topic while also providing methods for applying that knowledge within the context of different professional communication courses.

The book's fourth and final section represents a more general perspective on the educational context in which these ideas on culture and communication are taught. That is, the chapters in this concluding section provide more holistic approaches educators can consider when teaching the topics on culture and communication covered in the previous three sections of the book. In this way, the entries in this final section provide readers with mechanisms for combining the ideas and approaches noted in the book's earlier chapters into the context of a single class or an overall curriculum. That said, let's examine the focus and contents of the major section in this text.

The Contexts Examined in This Collection

The presentation of different context-related areas has been arranged to move gradually from more focused and specific items to broader ones. The idea is that the latter sections of the book address larger categories into which preceding sections can fit. Through this structure, readers can gradually expand their knowledge of international communication and see how different topics are connected to or are nested within others.

Design Contexts

At a foundational level, effective communication is about design. That is, the way in which a message is crafted—both in terms of verbal and visual content—has profound consequences for how it is received by different audiences. For this reason, the collection begins with the section "Design Contexts," and the entries in this section introduce readers to different verbal (i.e., linguistic) and visual (i.e., image-related) aspects individuals need to consider when designing messages for international audiences.

This initial section's first chapter, Maylath and Hammer's "The Imperative of Teaching Linguistics to Twenty-First-Century Professional Communicators," introduces readers to the ways in which language—and language-based practices such as translation—creates and affects the context in which international exchanges occur. In examining these ideas, Maylath and Hammer note how an understanding of linguistics can help professional communicators navigate different aspects of an international interaction. The authors also examine how professional communicators can draw upon a knowledge of linguistics to work more effectively with machine translation software, controlled language, single-sourcing, and content management. In this way, the chapter's authors introduce readers to verbal/linguistic factors to consider when designing communiqués or materials for individuals from different nations and cultures.

The second and third entries in this initial section expand upon this notion of design to include visual communication and the balance between visual and verbal communication in international settings. In the section's second chapter, "Cultural Contexts in Document Design," Yiqin Wang and Dan Wang use a comparative approach to examine how culture can affect design preference within genres—particularly in relation to balancing the uses of verbal and visual channels of communication. The chapter reports on instances where individuals from two different cultures were asked to review and comment on aspects of visual/verbal design employed in parallel technical documents designed for different cultural audiences. Wang and Wang use the results of these situations to note how culture can influence the overall design expectations individuals from different cultures associate with certain genres or documents. In so doing, the authors also note how cultural expectations create a context that individuals often rely on to assess, interpret, and use informational materials. Wang and Wang then conclude by overviewing exercises educators can use to introduce professional communication students to these ideas.

Audrey Bennett's "Teaching Image Standards in a Post-Globalization Age" takes this approach to teaching culture and design expectations a step further by presenting metrics educators can use to assess student work in this area. In her chapter, Bennett notes how factors of design and consumption—in combination with forces of globalization—have changed the context in which communication materials—particularly visuals—are now created, accessed, and interpreted/used. She also explains how such changes need to be integrated into current professional communication practices to reflect the expectations of the modern globalized workplace and related markets. To help readers better understand (and teach) these ideas, Bennett presents a framework readers can use when trying to create "communicatively effective images" for the global marketplace. She then discusses how these frameworks can become guidelines for teaching students about such issues and metrics for assessing student projects involving culture and design.

Societal Contexts

While language and design can create new situations for sharing ideas, both factors occur within greater societal contexts. It is these societal contexts that often create the ideas—and even the rules—individuals apply when using linguistic or visual approaches to share information. As a result, today's professional communicators need to understand how the societal contexts of different cultures can affect communication expectations. Professional communicators also need to understand and anticipate factors that can cause change in such societal contexts. While the variables involved are manifold, two areas that are particularly influential in relation to culture and communication are *technology* (i.e., the tools societies use to communicate) and *ethics* (i.e., the frameworks that tend to govern many communication practices in a given culture). For this reason, the entries in the book's second major section, "Societal Contexts," focuses on these items.

The second section starts with Brian Ballentine's chapter "Linux on the Education Desktop: Bringing the "Glocal" into the Technical Communication Classroom." In this entry, Ballentine examines how the growing global use of open source software (OSS) is changing how societies use computing technology to communicate. As he explains, the inexpensive and open nature of OSS has prompted a growing number of countries and cultures to use it for a variety of purposes. Within this context, education has been an area in which OSS adoption and use has grown steadily across the globe. However, as Ballentine points out, the cultural context in which OSS is employed often governs how it is used and who tends to use it. Thus, different national and cultural approaches to and perspectives on OSS use can create interesting situations related to international communication. It is therefore important that professional communication students familiarize themselves with OSS—and with the societal factors that affect its adoption and use—if they wish to use technology to communicate effectively in today's global workplace.

The next entry, Dan Voss and Bethany Aguad's "Teaching the Ethics of Intercultural Communication" reviews issues of social context, culture, and communication from an even higher level. In their chapter, Voss and Aguad examine different cultural perspectives on and approaches to ethics—particularly in terms of professional communication practices. In so doing, the authors provide approaches to navigating what they identify as the "thorny thicket" of ethical expectations in global contexts. They do so by overviewing different methods individuals can use to identify and understand the varying systems of ethics one may encounter internationally. Voss and Aguad also explain how these methods can double as mechanisms educators can use to teach this topic to professional communication students.

Online Contexts

Online media are continually changing the contexts in which international exchanges take place. Now, individuals on opposite ends of the globe can communicate almost as quickly and easily as persons working in the same office. As a result, online environments have created new and evolving contexts that bring with them a range of factors affecting professional communication practices. The challenge for educators becomes providing students with experiences they can use to better understand interaction in such environments. The book's third major section ("Online Contexts") provides initial ideas and approaches for achieving this objective.

In the first entry in this section, César Correa Arias overviews how online media are affecting educational practices in global contexts. Correa's chapter—"Autonomous

Learning and New Possibilities for Intercultural Communication in Online Higher Education in Mexico"—notes that the decentralization and interconnectedness of online media can create a new context for considering educational practices. In this context, students from anywhere in the world can participate in classes. Educators working in such environments therefore need to explore new methods for teaching students the professional communication skills essential for success in such settings. To examine these ideas, Correa draws from his own experiences teaching online courses at the University of Guadalajara, Mexico. He also provides suggestions for how educators might rethink pedagogical approaches to account for aspects of culture and cyberspace when teaching online.

The second entry in this section—Darina M. Slattery and Yvonne Cleary's "E-Learning and Technical Communication for International Audiences"—builds upon a number of Correa's earlier ideas by exemplifying how such concepts can be implemented in relation to teaching professional communication online. In their chapter, Slattery and Cleary examine the pedagogical strategies the faculty in the Technical Communication and E-Learning program at the University of Limerick, Ireland, have adopted to address aspects of culture and online media. The authors note that the faculty developed these strategies to help students gain skills in both global communication practices and e-learning methods. The overall approach involves using online media to allow students in Ireland to participate in virtual team projects with peers in the United States. In describing these activities, Slattery and Cleary connect ideas and approaches to various sources readers can use to better contextualize the practices presented in the chapter. In so doing, the authors provide readers with information and ideas on how to undertake similar international exercises via online media.

This third section concludes with Pamela Estes Brewer's "Teaching and Training with a Flexible Module for Global Virtual Teams," which provides a more detailed examination of using online media to provide students with international educational experiences. (Brewer's chapter thus provides a framework for applying many of the ideas advocated in the earlier chapters by Correa and by Slattery and Cleary.) Like the other authors in this section, Brewer draws from her own experiences to present approaches for achieving a particular educational objective (i.e., helping students develop effective professional communication skills related to international contexts). In particular, she focuses on how different global virtual team project (i.e., the use of online media to organize students in different nations into project-based teams) can introduce professional communication students to ideas of culture, communication, and cyberspace. Brewer also explains how she has used such projects to develop a flexible module other educators can use to engage in similar practices. She then provides examples of how she has used this module to engage in effective interactions involving student teams from a number of cultures.

Educational Contexts

The chapters in the previous major section offer suggestions on the kinds of topics to cover when teaching international issues in professional communication. In so doing, the authors of these chapters also provide suggestions for how to teach such topics and how to include different media in such practices. The question becomes how to integrate these ideas and approaches into professional communication classes and across curricula. Moreover, when thinking on both the course and the curricular level, educators need to consider assessment practices—or mechanisms for evaluating the success with which students learn such ideas and related communication skills. The entries in this final section represent approaches for integrating the teaching of intercultural communication into a range of contexts from the individual course to an overall curriculum. Also covered in this section are mechanisms educators can use to assess learning in these different instructional contexts. By covering these topics, this final section provides readers with a roadmap for integrating ideas covered in preceding sections into different instructional contexts.

This final section begins with Helen Grady's chapter "Strategies for Developing International Professional Communication Products." In this entry, Grady examines the merger of media and culture in terms of how to bring the two into the structure of an individual class. As Grady explains, a central aspect of training students for the global workplace is to provide them with the skills needed to collaborate on design practices distributed across nations. In discussing these ideas, Grady provides an example of how she met such an objective within the context of a particular online class. She uses these experiences to present strategies for teaching students about how different cultural factors shape the communication process in globally distributed design practices. In so doing, Grady includes example assignments designed to foster the knowledge base and skill sets needed to work in such contexts. She also provides sample student responses to those assignments to provide readers with a context for understanding how students might react within a similar class situation.

The section's second entry—Han Yu's chapter "Teaching Cultural Heuristics through Narratives: A Transdisciplinary Approach"—applies a more macro-level perspective on the ideas discussed by Grady. Instead of examining the teaching of culture and communication within the context of a specific class, Yu presents an overall framework—or a heuristic—individuals can use to integrate the teaching of culture and communication both into individual classes and across the classes in an overall curriculum. Yu's method, which focuses on teaching cultural heuristics/models through narratives, combines the functionalist and interpretive approaches to understanding culture. In this way, her framework allows educators to address and to integrate different cultural contexts (e.g., linguistic, design, societal, and online) in a more unified way when teaching students about international communication. In discussing her approach, Yu notes how it overlaps with and can be used in parallel with more traditional perspectives on teaching culture and communication.

Within the overall discussion of ideas and approaches to education, one question remains: "How do we assess the success of student learning?" An answer to this question is provided in the section's final entry, Darla K. Deardorff and Duane L. Deardorff's "Assessing Intercultural Outcomes in Engineering Programs." In this chapter, Deardorff and Deardorff provide readers with ideas on the assessment of intercultural learning outcomes within engineering programs. They do so by examining different frameworks educators (and others) can use to do assessments of classes (or curricula) that contain cultural components. In presenting these ideas, Deardorff and Deardorff examine principles associated with effective assessment practices and offer examples of intercultural outcomes assessment. They also share a checklist readers can use to develop assessment plans for their own classes and programs.

Conclusion

A famous proverb notes "A journey of a thousand miles begins with a single step." This expression perhaps best describes the purpose of this collection (a single step) within the context of the overall topic it examines (educational practices in international professional communication).

> The editors believe this collection provides readers with an effective initial foundation for understanding both topics to teach and approaches to use when integrating ideas of culture into professional communication classes.

The structure of the text, in turn, moves readers through progressive levels of understanding aspects of this overall topic within different contexts—from that of a specific item within international communication to the assessment of teaching multiple aspects of culture and communication across classes and curricula. Ideally, this structure will allow readers to examine these ideas at a pace and a level that best suits their backgrounds and objectives in relation to this subject.

The editors also hope that the ideas presented in this volume might serve as a mechanism for prompting further examination relating to teaching international professional communication. For this reason, the editors encourage readers to consider how they can expand on perspectives and approaches provided in the text as well as perhaps conduct future research related to teaching this topic and assessing related outcomes. In sum, the journey related to understanding—and teaching—professional communication in international contexts is one in which we—as educators—can all travel together. It is also a context in which we—as educators—can benefit from the experiences of others traveling this same path. It will be interesting to learn where the next steps in this journey take us…

Kirk St.Amant
Louisiana Tech University and University of Limerick

Design Contexts

1

The Imperative of Teaching Linguistics to Twenty-First-Century Professional Communicators

Bruce Maylath

North Dakota State University

Steven Hammer

Saint Joseph's University

Because professional communication in the twenty-first-century so often includes language translation, and because the costs of translation have fostered the growth of high-speed, efficient translation tools, professional communicators increasingly need to have a conscious and comprehensive knowledge of grammatical structures to use such tools. This chapter examines how professional communicators must apply a thorough knowledge of linguistics when they employ software for machine translation, controlled language and terminology management, structured/guided authoring, and single-sourcing and content management. It identifies the most important syntactical structures and linguistic features for professional communicators to understand. It also demonstrates the importance of mastering at least one language beyond one's native language.[1]

---

[1] All student names have been changed to protect privacy.

Teaching and Training for Global Engineering: Perspectives on Culture and Professional Communication Practices, First Edition. Edited by Kirk St.Amant and Madelyn Flammia.
© 2016 The Institute of Electrical and Electronics Engineers, Inc. Published 2016 by John Wiley & Sons, Inc.

Note: In this chapter, the authors describe their experiences working together on the Trans-Atlantic Project. Hammer began involvement in the project as Maylath's student, and later continued as an instructor and colleague.

1.1 Introduction

Outside of native English-speaking countries, particularly the United States, it is rare to find an engineer who would not find mastery of more than one spoken and written language vital to have a successful engineering career. (When we say "language," we are not referring to mathematical language or computer programming languages but human languages.) In a global economy, engineers commonly work on projects beyond their national borders. Even if they are not required to travel to a project in another country, they contribute to designs as members of cross-cultural virtual teams (CCVTs). Frequently, an engineer can communicate effectively in his or her native language plus English, which, as linguist David Crystal [1] has demonstrated, has effectively become the first language in world history to achieve the status of a truly global language.

The situation in the United States, however—and to some degree in the United Kingdom and Australia as well—is quite different among engineers. As the offspring of a family of civil engineers, one of us (Bruce) remembers well the discussion among engineers in the 1960s about the decisions being made at US engineering colleges to eliminate the foreign language requirement, thereby leading a trend that colleges in other disciplines, such as agriculture and pharmacy, would imitate. The argument behind the decision at that time was that engineers rarely worked outside the United States and therefore did not need any language besides English. A global marketplace and World Wide Web were never envisioned.

Worse, the US engineering colleges' decisions to dispense with the foreign language requirement reduced the demand for college-bound high school students to learn a foreign language in high school. An unintended consequence is that foreign language offerings—and, indeed, entire language departments—have been slashed over the past few decades all across America's high schools and state colleges and universities [2]. By 2012, only 37% of all universities and colleges in the United States had any requirement for undergraduates to learn a foreign language [3].

Heightening linguistic knowledge and language awareness among engineers and professional communicators in the engineering field—especially among those individuals whose first language is English—is a key ingredient to successful communication in a twenty-first century global working environment. This chapter aims to demonstrate

- Why language awareness and some linguistic knowledge are crucial competencies for professional communicators in a variety of fields
- How the study of language was hollowed out of English classes even as foreign languages were being cut
- What engineering and professional communication students need to know about linguistics and language.

Through examining these topics, professional communicators can develop an appreciation for learning a second language, increase their awareness of language, and begin to apply a knowledge of linguistic features and grammatical structures when they craft written texts to be used outside of their own culture and/or language.

1.2 Why Linguistics? The Dangers of Monolingualism

The attitude that one language, namely English, suffices for the communication needs of twenty-first century professional engineers is a dangerous fallacy. The many peoples whom today's century professional engineers serve are simply not uniformly fluent in English [4]. Moreover, the proliferation of varieties of English, wherever it is learned, can result in lost understanding and misunderstanding when speakers of different varieties of English must communicate with each other, even when the varieties are widespread, well-known standards. (Our favorite example is the verb "to table" in deliberative settings; its meanings in North American and British English are exactly the opposite of each other.) The many "World Englishes," as Crystal [1] has dubbed them, require heightened language awareness among professionals, especially professional communicators. As Charles and Marschan-Piekkari [5] have warned, "It is not enough to learn to understand British/North American/native speaker English. MNC [Multinational Corporation] staff should be exposed to different 'World Englishes'" [5, p. 24]. And different languages, we might add.

The calls for language awareness and linguistic knowledge now come from every area of human enterprise. Swearingen, Barnes, Coe, Reinhardt, and Subrahanian [6] stress that "globalization of the manufacturing profession will require its practitioners to master engineering methodologies, cultures, and languages from more than one country" [6, p. 256]. Downey et al. [7] call for "globally competent engineers" [7, p. 13] who can, among other competences, deal with differences in language. Lohmann, Rollins, and Hoey [8] observe that "the facility to communicate in other languages and to assimilate with ease into foreign workplaces and lifestyles are critical to both professional and life success" [8, p. 121]. The authors list five elements of global competence, first among them "proficiency in a second language" [8, p. 128].

Engineering students can be loath to acknowledge the importance of proficiency in language, even one's first language, much less a second. Bruce recalls asking students in his first class of Engineering Communication many years ago why they had chosen to study engineering. The majority response: "To avoid writing." Yet as any veteran engineer would tell students, clear writing, as well as speaking, is critical to success in engineering.

Bruce recalls asking students in his first class of Engineering Communication many years ago why they had chosen to study engineering. The majority response: "To avoid writing." Yet as any veteran engineer would tell students, clear writing, as well as speaking, is critical to success in engineering.

Even for those competent in writing in their first language, not recognizing others' languages can cost lives. Reinhard Schäler [9], director of the University of Limerick's Localisation Research Centre, calculates "that more than the cost of multilingualism is the cost of the lack of multilingualism" [9, p. 222]. Schäler gives examples from healthcare and economic development of people dying simply because information came to them in a language that was not their own, instead of "in the language they need...to make appropriate decisions and save lives" [9, p. 223]. Schäler concludes, "...as long as access to vital and potentially life-saving online health information in one's native language is not made available, tens of thousands of people will continue to face certain death every day" [9, p. 224].

Engineers whose first language is English, especially in the United States, have not gained a stellar reputation in acknowledging, much less acting on, the fact that the users of their technologies and designs are not monolingual. In one of the most widely publicized such instances, an earlier decision by North American software engineers forced the German Language Council in 1999 to eliminate the centuries-old letter ß from the German alphabet. What caused this change? In creating URLs for Websites, the North American engineers were not even aware that languages can and often do have more than the 26 letters used in English. As a result, they developed a system that did not permit users to add characters beyond the 26 letters that exist in the English alphabet. Though hardly life-threatening, the German Language Council's bending to the decision of engineers who assumed English is all that matters bespeaks the power of ignorance and arrogance and the Law of Unintended Consequences, namely that lack of language awareness and suppositions about one's local conditions being the norm everywhere, can wreak untold harm on others beyond one's locale.

Today technical information must often be simultaneously translated and localized into 40 or more languages. Thanks to new trade agreements in the 1990s, such as NAFTA and WTO, Marschan, Welch, and Welch [10] would point out that language is the key factor in managing multinational businesses. They noted as well that power is concentrated in the hands of those who decide which language will be used. Much more recently, Zemliansky and Kampf [11] have observed that "...the American-centric view of the use of English in professional communication settings may not be sustainable any more" [11, p. 223]. "Incorporating international preparation into engineering curricula, however, has proven to be a major challenge," admit Lohmann, Rollins, and Hoey [8]. "It is due largely to the highly sequenced and content-demanding nature of the curriculum. Nonetheless, engineering programs are finding ways to incorporate language preparation" [8, p. 119].

1.3 Linguistic Knowledge—Where Did It Go?

Forty years ago, not only were foreign language requirements and offerings more abundant in the United States, so was attention to language and linguistics in English classes. It may surprise older readers that English classes in most North American schools (including universities) devote little attention to the study of grammar, semantics, pragmatics, and other sub-areas of linguistics—unless, that is, they've ever tried to talk to younger generations about theses topics, especially as they pertain to writing. It will come as even more of a surprise that even Ph.D. students in English—most of whom are themselves teaching writing as graduate assistants—have gained little exposure in grade school, high school, or undergraduate courses to principles and terms in linguistics, beyond parts of speech.

As an example, Steven offers his own experience. Until he took Bruce's graduate courses in History of the English Language and International Technical Writing (Steven's experience in the latter is described in Maylath et al., [12]), he had never heard a teacher explain such basic grammatical features and terms as gerunds and pronoun antecedents. Now, as he teaches advanced writing courses for junior and senior undergraduates to prepare them for writing in their professions, he recognizes the value of helping his undergraduates master linguistic concepts and terms. Throughout the remainder of this chapter, we offer examples from two of Steven's recent writing students to show how the students in these classes benefited from such instruction in linguistics.

How did the US English curriculum reach this point? Horner and Trimbur [13] show how the institutionalization of freshman English courses at US colleges and universities, starting with English A at Harvard in 1870, began to establish "a language policy that replaced the bilingualism (in principle if not always in practice) of the classical curriculum with a unidirectional monolingualism" [13, p. 595]. In the United States, knowing another language better than knowing English was viewed as a problem caused by "a deficit of immigrants," solved by eliminating the immigrants' languages [13, p. 617] along with Native American languages [14, p. 309]. Then, shortly after engineering colleges in the United States began eliminating foreign language requirements in the 1960s, attention to language, including grammar, was gradually expunged, over a 40-year span, from English teaching in America. In her article "The Erasure of Language," Peck MacDonald [15] delineates the how and why of this move toward monolingualism. Chief among the culprits that she identifies are the English education programs in teacher colleges that convince "new teachers-in-training who spent their early years in school assuming grammar need not be taught and would only be hurtful if it were" that their assumptions are correct [15, p. 605], even in the face of evidence, such as that collected in Mulroy [16], that demonstrate the opposite over time.

As Horner, NeCamp, and Donahue [17] make clear, problems result from

> restrictions English monolingualism has imposed on many of those…schooled in the United States: our schooling in language other than English, and official incentives to pursue the study of languages other than English, have been limited. U.S. students who do grow up bilingual or multilingual often find little support in school for those abilities. [17, p. 277]

Such restrictions have created a situation in the United States, where "…writers, speakers, and readers are expected to use Standard English or Edited American English—imagined ideally as uniform—to the exclusion of other languages and language variations"…under the assumption "that heterogeneity in language impedes communication and meaning" [14, p. 303].

Yet, in a global work environment involving CCVTs whose home languages number easily in the hundreds, such imagining of uniformity as ideal is far from realistic or beneficial. As Charles and Marschan-Piekkari [5] found, "native speakers of English might well be seen to be external to the problem. However, as our Kone interviews showed, they, too, both have and cause communication problems" [5, p. 25]. Indeed, "communication, or the lack of it, and language skills, or the lack of them, had implications particularly for power issues and efforts to create a feeling of togetherness, a unified and shared corporate culture. Language skills thus acquired significant strategic proportions" [5, p. 26]. In short, monolingual native speakers may think that they have communicated, but their lack of linguistic knowledge and language awareness too often prevents them from realizing that they have not. The next section delineates the advantages that accrue to those who learn another language and heighten their awareness of language(s).

1.4 Linguistics for the Professional Engineer and Communicator

Mastering a second (or third or fourth) language has obvious benefits when communicating with another professional who has a command of that language. In addition, merely having studied another language allows one to recognize ambiguities and opacities when using one's own language with others for whom it is not native. For instance, a North American who has grappled with gaining a command of, say, Spanish is more likely to empathize with a Korean attempting to communicate in English. Consequently, the North American is more likely to feel compelled to sustain efforts at pursuing clarity and comprehension. Most importantly, study of a second language—any language—to at least an intermediate level allows one to become aware that translation between languages is not a matter of simply substituting the words of one language for those of another. Rather, translation is a matter of attempting to find equivalent means of expressing ideas in different languages and cultural contexts.

Becoming conscious of this reality is part of what linguists and language educators refer to as "language awareness." Now a worldwide movement among educators—one with an eponymous professional association and journal (see http://www.language awareness.org/ language awareness has been part of the English curriculum in the United Kingdom since the 1990s. As its goal, "Language Awareness seeks to move students from intuitive powers to conscious awareness of how users create meaning through language" [18, p. 3]. When this goal is achieved and put into practice, two of four functions of language, as identified by Tulasiewicz [19], are heightened by language awareness:

1. Promoting general empowerment by providing language users with a sound command of their language(s) and enabling them to be efficient in all negotiations involving language;
2. Activating the instrumental function of language to facilitate communication, if necessary in more than one language. [19, p. 8]

Such functions are at the core of what professional communicators must enact daily in order to operate in multilingual and multicultural global work environments, including CCVTs. As Louhiala-Salminen and Kankaanranta [20] put it,

First, communicators need to be prepared and equipped to check for understanding and make clarifying questions. Second, since communicators need to show tolerance toward different Englishes, they should ideally receive some training in identifying and adjusting to different pronunciations, accents, and intonations of English. Third, they should be aware of and flexible toward the various mother tongue discourse practices reflected in EFL [English as a Foreign Language] communication. [20, p. 260]

These points are aimed toward interlocutors using only English as a *lingua franca*. Language awareness and more specific linguistic knowledge are even more critical when writing and preparing a technical document for translation.

1.4.1 Linguistics for Translation

Increasingly, professional communicators are involved in translation, first and foremost as project managers of their companies' or agencies' translation projects. In addition, a trend has emerged of professional communicators seeking cross-training as translators, and vice versa, a convergence that has accelerated rapidly in Europe but which is also taking place in North America [21]. For professional communicators writing and preparing texts for translation, it is essential to understand grammatical structures, lexical features, semantics, and pragmatics. Doing so aligns with the grammatical, sociolinguistic, discursive, strategic competences conceived by Canale and Swain (1980) and verified in our own research of technical writing and translation students cooperating in CCVTs [22, 23].

In many advanced writing courses, such as North Dakota State University's Writing in the Technical Professions (English 321), we have taught many engineering students to identify, name, and manipulate specific linguistic features. The aim of this activity is to improve the text in order to make the translators' job easier, using Maylath [24] as a guide to prepare the text for translation. A sample follows with extracts taken by permission from instructions written and prepared for translation by one of Bruce's former students:

Substituting infinitives for gerunds
1. Objective: Achieving greater speed with less effort is the main purpose of the digitizer.
2. Objective: To achieve greater speed with less effort is the main purpose of the digitizer.

Example: Inserting the relative conjunction "that"
1. Notice there are no rollers on the bottom of the puck.
2. Notice that there are no rollers on the bottom of the puck.

Example: Clarifying unclear antecedents
1. Electronic pulses start within the tablet and locate the cross-hairs on the puck to determine its location. . . . Notice the resemblance between the mouse and the puck, when it's within the screen-pointing area.

2. Electronic pulses start within the tablet and locate the cross-hairs on the puck to determine the puck's location. . . . Notice again the resemblance between the mouse and the puck, when the cross-hairs are within the screen-pointing area. [24, pp. 349–350]

Preparing a text in these ways might seem less than complicated to those well versed in grammar and writing instruction from decades long past. However, as Steven discovered in his first translation project as a student, deprived until his graduate studies of such instruction, the terms themselves (gerunds, antecedents, etc.) design an anxiety-provoking blur.

The initial problem is grammatical nomenclature: jargon terms like "infinitive," "gerund," "conjunction," and "antecedent" are all Greek to English speakers, or, more accurately, are Greco-Latinate in origin, and thus opaque and foreign [25, 26]. However, those are the terms that we are stuck with in English when discussing the operations and the nuances of language. As Klammer, Schulz, and Della Volpe [27] observe, "linguists must invent language to describe language" [27, p. 88]. Thanks to Latin's history as a *lingua franca* before English spanned the globe, the nomenclature of English grammatical terms remains Greco-Latinate and shows no signs of changing. As much as North American students might chafe and rebel against them, they are essential to know in order to talk to other authoring and translation CCVT members about the texts that they are producing.

Just how this is so emerged in comments from students in Steven's advanced writing courses when, in 2011, he joined the long-running Trans-Atlantic & Pacific Project as an instructor. In keeping with the Trans-Atlantic & Pacific Project's long-time goal of creating collaborations to resemble many professional communication workplaces, Steven and his European partners formed CCVTs comprising writing students in North America and translation students in Europe. (For more on the Trans-Atlantic & Pacific Project and how it operates, see [23] and [28–30].) A writing student whom we'll identify as "Emily" expressed some insecurities while getting started with the project due to a lack of linguistic knowledge. (All student comments presented in this paper are reproduced with permission of the individual):

> An important aspect of this portfolio was to prepare a paper that I wrote for translation by looking for specific language, gerunds, and antecedents. Personally, I have never worked with these specific parts of sentences; therefore, it was rather difficult for me to pick out the gerunds and antecedents. I have always struggled with grammar, which make [sic] this assignment especially difficult for me.
>
> Another part I struggled with was knowing how many pronouns to change, because I did not want to change everything because that might be insulting to my translator. I did not want them to think that I thought they could not link pronouns back to the previous sentence and still understand what is going on in the sentence. Also, I did not know what to put for my glossary, because again I did not want to insult the intelligence of any of my partners, but I also did not know how in-depth their knowledge of the English language was.

Inserting direct instruction of linguistics, grammar, and accompanying terminology into the writing course proved its usefulness. Several students commented that the linguistic lessons in class were helpful (and quite necessary) in aiding their ability to carry out the project. Another student, "John," said,

> I will definitely save the 'Preparing for Translation' document the class received from Dr. Maylath in case I have to do this type of work in the future. The tips on this document were validated to me when I missed one of the gerunds in my paper. My Belgium [sic] partner asked me about the phrase:
> "Humans try to add meaning to their life, to become noble and important to others…"
> She asked if I meant:
> "to become noble to others and important to others"
> Or-
> "They try to become noble. They try to be important to others."
> Although both statements make sense, it could have been easily clarified and sounded more professional if I had removed the gerund from my text and changed the phrase to:
> "Humans try to add meaning to their life, to become noble to others and to become important to others…"

These students' observations not only reflect their acquisition of linguistic concepts and terms but also their growing awareness of ambiguity or clarity as texts move between languages.

1.4.2 Who Is Ready; Who Is Not

In nearly all of our advanced writing courses, we find that few US students have studied a foreign language for more than two years in high school. In the case of engineering students, as a result of the decision by engineering colleges to banish their foreign language requirements in the 1960s, it is not unusual that students have taken no foreign language whatsoever. (The accrediting association ABET, which establishes criteria and program reviews for engineering degrees, lacks any criteria for learning foreign languages, though it does include "an ability to apply written, oral, and graphical communication in both technical and non-technical environments" as a desired student outcome [31].) Thus, when faced with a writing assignment involving translation preparation, followed by a working partnership with a European peer who has a significantly greater conscious knowledge of the writers' native language than the writers do, the writers' confidence drops precipitously and affects their performance in the project. North American student "Jennifer" remarked,

> This is an interesting project because very few of us [North] American students even speak a second language. The grammar of another language is very foreign, to me at least, so this has been very interesting to think about the structural and idiomatic differences between our languages.

Among those students who were not fluent in a second language, several comments reveal that even introductory courses in foreign languages were useful to them in understanding grammatical features and differences between English and their partners' native languages (the target languages for their documents), even if they overgeneralize in their hopes that other target languages are the same as the one they have studied. Student "Mark" ruminated,

> [R]ecalling high school French class, I tried to use the appropriate convention of sentence structure, reversing the order of word arrangement at times in a way that made translation easier from French to English, with my fingers crossed that the same basic rules applied to Dutch.

"John" also applied such training to a new language and situation: "I'm currently enrolled in a beginner level German class, and I've just recently learned you can use the word 'und' (and) as many times in a sentence as you like. I assume it is like this for many languages including Italian and Dutch."

Additionally, some students perceived how this project could aid them in future professional situations, even those whose professions might not involve document translation. "John" wisely reflected,

> I also believe that this experience of learning some of these translation tools will help me in my profession as a teacher as I deal with the ever increasing population of English Language Learner students. Knowing how to clarify language will hopefully help me be a better educator to the students who, like my transatlantic partners, do not know English as a first language. By previewing texts we are going to read or assignments we are going to do in class, I could clarify unclear language, locate idioms, and identify words that may be pointed out to the student in a glossary or verbally.

In short, learning key elements of linguistics proved imperative for these budding professional communicators. Table 1.1 sums up the language instruction and experiences, or lack of them, that make professional communication in the twenty-first century global economy smooth or rough for its practitioners.

1.4.3 Beyond the Classroom

Although the courses from which we have drawn examples are categorized as advanced writing, the truth is that they can give students only a taste of the complexity of professional communication to come in their future workplaces. For the professional engineer and communicator who regularly work with texts that require translation, the need to employ linguistic knowledge runs much deeper. This situation is largely due to the demands of language translation software, like Trados, which requires a knowledge of grammatical structures for users to be able to mark phrases that can be reused safely without requiring a second translation—and resulting additional costs.

TABLE 1.1. Summation of Language Instruction Experiences

Heightened Language Awareness	Lowered Language Awareness
Lessons in grammar in grades K-12	Few lessons in grammar in grades K-12
Attention to grammar in college writing courses	Inattention to grammar in college writing courses
Learning other languages, particularly their grammar	Avoiding learning other languages and particularly their grammar
Taking linguistics courses, especially Intro to Linguistics or Grammatical Structures	Not having taken other linguistics courses, especially Intro to Linguistics or Grammatical Structures
Tutoring in a writing center	Not having tutoring experience in a writing center
Reading grammar handbooks	Not reading grammar handbooks

Current translation processes typically involve initial machine translation, structured/guided authoring, and single-sourcing and content management. To use such software packages, professional communicators need to understand terminology management, such as making sure that only a single term—and no synonyms—are used to name a technical feature, and controlled language, such as limiting each sentence to a single idea. When used effectively, these methods save authoring and translating time and costs exponentially. They are the key to using and reusing content/text across documents, as anyone familiar with single-sourcing or content management can attest.

For example, Spyridakis, Holmback, and Shubert [32] were among the earliest researchers to demonstrate the advantages of using simplified English and controlled language in professional communication practices (in their case, with Boeing's manuals for jet airplanes and their machinery; see http://www.boeing.com/phantom/sechecker/se.html). In more recent years, Uwe Muegge [33, 34], who pioneered groundbreaking terminology management systems for the medical device engineering firm Medtronic, has emphasized how a thorough knowledge of linguistics, particularly lexical and grammatical structures, is essential to the efficacy and efficiency of technical documentation that will be translated into multiple languages. "Terminology management is the most efficient solution," writes Muegge, "for ensuring that the organization as a whole uses the same terms to describe the same features and functions" [35, p. 16]. The exposure that our advanced writing students receive to linguistics through direct instruction and the experiential Trans-Atlantic & Pacific Project is basic: for example, that pronouns refer to antecedent nouns or that gerunds can be recast as infinitive phrases. However, Bruce has had alumni—especially engineers—tell him that such teaching and learning put them head and shoulders above their North American co-workers when handling translation projects.

1.4.4 Sociolinguistics and Pragmatics

Earlier, we mentioned Canale and Swain's [36] listing of sociolinguistics and discourse as necessary linguistic competencies, and then went on to explain how our own study of linguistic pragmatics [23] verified their work. Our research on linguistic pragmatics as practiced between writing and translation students revealed both subtle power moves and attempts at diplomacy through the manipulation of language in e-mail messages as the students negotiated meanings from the source texts being rendered in the target languages. Such competencies are often overlooked in English or in writing courses, even though these competencies are essential to effective communication. For example, in a section titled "Interactional Sociolinguistics" of their book chapter "Intercultural Professional Communication," Kong and Cheng [37] describe miscommunication among employees because Indian English prosody, especially syllable stress patterns, differ from the prosodies of other Englishes. They also describe frustration with Chinese users of English because they "tend to delay the main point of their arguments by using a lot of prefacing strategies such as giving a number of reasons before the main point" [37, p. 9]. Examining another area of sociolinguistics and pragmatics, a study of workplace communication by Ladegaard [38] revealed that humor

> is used to maintain good relationships among colleagues, as a way of "doing collegiality," but it is also used extensively as a way to "do power" less explicitly—a generally more acceptable strategy to challenge existing power relationships in a context where informality is perceived as valuable and status differences played down. [38, p. 207]

In our own classes, many students now use several social media services, especially Facebook®. While this medium of communication often yields little in terms of project progress, it does result in more frequent interactions between collaborators, which in turn results in more collegial relations. Several students have noted that they found such contact to be not only enjoyable socially but also useful as they and their partners became more approachable and familiar to each other. "Chris" wrote that

> Elena and I mainly communicated with email and some Facebook. When I did talk to Elena on Facebook, it was mostly conversations about what we had in common. Elena was more interested in getting to know me as a person instead of helping me with my translation project, which in itself is not a bad thing. This type of discussion opened a line of communication that I would have never had without this class... I communicated with Cedric through Facebook and emails on average every other every day or every few days. At the least, we communicated three times a week.

"Chris" echoed this experience again:

> My main interactions were through email which I found most productive. The few interactions between Facebook and Skype shown were mainly personal "get to know you" exchanges in order to have better relationships with my partners. I was able to learn many pieces of both Belgium [sic] and Italian culture through these exchanges; for example the Belgium beliefs and experiences

of Halloween. I believe that a better working relationship begins with better communication and healthier relationships. I understand that using a second language with native speakers is a hard task. To make my partners feel more welcomed and at ease I wrote to them with their native languages.

"Ashley" also shared this experience:

In the beginning, we emailed back and forth almost every week to maintain communication and to be available for any questions that we had for each other. After emailing frequently with one another, we became more comfortable with each other. Before I knew it I had two new friend requests on Facebook, from both my Italian and Belgium partner. Facebook is a popular social networking website. It was nice to be able to utilize this website and put a face with the names of my two partners. After communicating on Facebook approximately five times we began to get to know one another and became much more comfortable. I even got the courage to ask both my partners to Skype with me.

Through such exchanges, and discussions about them in class, we have been able to realize Horner, Lu, Jones Royster, and Trimbur's goal [14]: to "encourage renewed focus by students of writing on the problematics of translation to better understand and participate in negotiations of difference in and through language…" [14, p. 308]. Our students at NDSU echo Wittman and Windon's [39] student, who, after finishing an English course in which exposure to texts in translation was emphasized: "Overall, I think I have learned more between languages than I ever could in just one language" [39, p. 452].

1.5 Conclusion

Wittman and Windon conclude that "…a translation studies course should…be mandatory for all undergraduate English majors" [39, p. 449]. We do not see that option as a viable one for engineering students, though such may be possible for those earning a minor in technical writing. Rather, we agree with Horner and Trimbur [13] that "the task, as we see it, is to develop an internationalist perspective capable of understanding the study and teaching of written English in relation to other languages and to the dynamics of globalization" [13, p. 624]. Our, and our colleagues', Trans-Atlantic & Pacific Projects provide a highly effective, low-cost model for inserting linguistic instruction and relevant experiential practice into existing writing courses required for professional engineering and communication students.

Throughout this chapter, we have portrayed the crying need for professional communicators, first and foremost in the engineering fields, to understand and employ a basic knowledge of linguistics. We have shown how, in the United States, curricula that formerly included linguistics and foreign language instruction while heightening language awareness have been allowed to let such instruction fade away to nearly nothing. Finally, through the Trans-Atlantic & Pacific Project, we have provided methods and a model for bringing such instruction back in ways that have proved highly relevant and successful to students in college-level writing courses. We encourage others to follow our lead.

References

1. D. Crystal, *English as a Global Language*, 2nd ed. Cambridge: Cambridge University Press, 2003.
2. N. Furman, D. Goldberg, and N. Lusin. *Enrollments in Languages Other Than English in United States Institutions of Higher Education, Fall 2009*. New York: Modern Language Association, 2010. Available at: http://www.mla.org/pdf/2009_enrollment_survey.pdf.
3. K. Fischer, "U.S. colleges' global efforts fall short in areas like languages and faculty travel," *The Chronicle of Higher Education*, p. A18, July 6, 2012.
4. V. Ginsburgh and S. Weber, *How Many Languages Do We Need? The Economics of Linguistic Diversity*. Princeton, NJ: Princeton University Press, 2011.
5. M. Charles and R. Marschan-Piekkari, "Language training for enhanced horizontal communication: A challenges for MNCs," *Business Communication Quarterly*, vol. 65, no. 2, pp. 9–29, 2002.
6. C. Swearengen, S. Barnes, S. Coe, C. Reinhardt, and K. Subrahanian, "Globalization and the undergraduate manufacturing engineering curriculum," *Journal of Engineering Education*, vol. 91, no. 2, pp. 255–261, 2002.
7. G. L. Downey, J. C. Lucena, B. M. Moskal, R. Parkhurst, R. T. Bigley, C. Hays, B. K. Jesiek, L. Kelly, J. Miller, S. Ruff, J. L. Lehr, and B. A. Nichols, "The globally competent engineer: Working effectively with people who define problems differently," *Journal of Engineering Education*, vol. 95, no. 2, pp. 1–17, 2006.
8. J. R. Lohmann, H. A. Rollins, and J. J. Hoey, "Defining, developing and assessing global competence in engineers," *European Journal of Engineering Education*, vol. 31, no. 1, pp. 119–131, 2006.
9. R. Schäler, "Information sharing across languages," in *Computer-Mediated Communication Across Cultures: International Interactions in Online Environments*, K. St.Amant, and S. Kelsey, Eds. Hershey, PA: IGI Global, 2012, pp. 215–232.
10. R. Marschan, D. Welch, and L. Welch, "Language: The forgotten factor in multinational management," *European Management Journal*, vol. 15, no. 5, pp. 591–598, 1997.
11. P. Zemliansky and C. Kampf., "New landscapes in professional communication: The practice and theory of our field outside the US," *IEEE Transactions on Professional Communication*, vol. 54, no. 3, pp. 221–224, 2011.
12. B. Maylath, S. Vandepitte, P. Minacori, S. Isohella, B. Mousten, and J. Humbley, "Managing complexity: A technical communication/translation case study in multilateral international collaboration," *Technical Communication Quarterly*, vol. 22, no. 1, 2013.
13. B. Horner and J. Trimbur, "English only and U.S. college composition," *College Composition & Communication*, vol. 53, no. 4, pp. 594–630, 2002.
14. B. Horner, M. Lu, J. Jones Royster, and J. Trimbur, "Language difference in writing: Toward a translingual approach," *College English*, vol. 73, pp. 303–321, 2011.
15. S. Peck MacDonald, "The erasure of language," *College Composition & Communication*, vol. 58, no. 4, pp. 585–625, 2007.

REFERENCES

16. D. Mulroy, *The War Against Grammar.* Portsmouth, NH: Boynton/Cook, 2003.
17. B. Horner, S. NeCamp, and C. Donahue, "Toward a multilingual composition scholarship: From English only to a translingual norm," *College Composition & Communication*, vol. 63, no. 2, pp. 269–300, 2011.
18. L. J. White, "Introduction," in *Language Awareness: A History and Implementations*, L. J. White, B. Maylath, A. Adams, and M. Couzijn, Eds. Amsterdam: Amsterdam University Press, 2000, pp. 1–3.
19. W. Tulasiewicz, "Whither language awareness? Enhancing the literacy of the language user," in *Language Awareness: A History and Implementations*, L. J. White, B. Maylath, A. Adams, and M. Couzijn, Eds. Amsterdam: Amsterdam University Press, 2000, pp. 5–20.
20. L. Louhiala-Salminen and A. Kankaanranta, "Professional communication in a global business context: The notion of global communicative competence," *IEEE Transactions on Professional Communication*, vol. 54, no. 3, pp. 244–262, 2011.
21. M. Gnecchi, B. Maylath, F. Scarpa, B. Mousten, and S. Vandepitte, "Field convergence between technical writers and technical translators: Consequences for training institutions," *IEEE Transactions on Professional Communication*, vol. 54, no. 2, pp. 168–184, 2011.
22. B. Mousten, S. Vandepitte, and B. Maylath, "Intercultural collaboration in the trans Atlantic project: Pedagogical theories and practices in teaching procedural instructions across cultural contexts," in *Designing Global Learning Environments: Visionary Partnerships, Policies, and Pedagogies*, D. Stärke-Meyerring and M. Wilson, Eds. Rotterdam: Sense Publishers, 2008, pp. 129–144.
23. B. Mousten, J. Humbley, B. Maylath, and S. Vandepitte, "Communicating pragmatics about content and culture in virtually mediated educational environments," in *Computer Mediated Communication across Cultures: International Interactions in Online Environments*, K. St.Amant and S. Kelsey, Eds. Hershey, PA: IGI Global, 2012, pp. 312–327.
24. B. Maylath, "Writing globally: Teaching the technical writing student to prepare documents for translation," *Journal of Business and Technical Communication*, vol. 11, no. 3, pp. 339–52, 1997.
25. D. Corson, *The Lexical Bar*, Riverside, NJ: Pergamon Press, 1985.
26. D. Corson, *Using English Words*. Dordrecht, The Netherlands: Kluwer Academic Publishers, 1995.
27. T. P. Klammer, M. R. Schulz, and A. Della Volpe, *Analyzing English Grammar*, 6th ed. New York: London, 2010.
28. J. Humbley, B. Maylath, B. Mousten, S. Vandepitte, and L. Veisblat, "Learning localization through trans-Atlantic collaboration," in *Proceedings of the IEEE International Professional Communication Conference,* 2005, pp. 578–595.
29. B. Maylath, S. Vandepitte, and B. Mousten, "Growing grassroots partnerships: Trans-Atlantic collaboration between American instructors and students of technical writing and European instructors and students of translation," in *Designing global learning environments: Visionary partnerships, policies, and pedagogies*, D. Stärke-Meyerring and M. Wilson, Eds. Rotterdam: Sense Publishers, 2008, pp. 52–66.
30. B. Mousten, B. Maylath, S. Vandepitte, and J. Humbley, "Learning localization through trans-Atlantic collaboration: Bridging the gap between professions," *IEEE Transactions on Professional Communication,* vol. 53, no. 4, pp. 401–411, 2010.
31. "Criteria for accrediting engineering technology programs, 2012–2013," ABET. Available at: http://www.abet.org/criteria-engineering-technology-2012-2013/.

32. J. H. Spyridakis, H. Holmback, and S. K. Shubert, "Measuring the translatability of Simplified English in procedural documents," *IEEE Transactions on Professional Communication,* vol. 40, no. 1, pp. 4–12, 1997.
33. U. Muegge, "Möglichkeiten für das Realisieren einer einfachen Kontrollierten Sprache [Implementing a simple controlled language]," *Lebende Sprachen,* vol. 3, pp.110–114, 2002.
34. U. Muegge, (2006). "Fully automatic high quality machine translation of restricted text: A case study,"in *Proceedings of the Twenty-Eighth International Conference on Translating and the Computer,* 2006, pp. 1–6.
35. U. Muegge, "Focus on: Terminology management," *BioProcess International,* vol. 4, pp. 16–20, 2010.
36. M. Canale and M. Swain, "Theoretical bases of communicative approaches to second language teaching and testing," *Applied Linguist,* vol. 1, no. 1, pp. 27–31, 1980.
37. K. C. C. Kong and W. Cheng, "Intercultural professional communication: Approaches and issues," in *Professional Communication: Collaboration Between Academics and Practitioners,* W. Cheng and K. C. C. Kong, Eds. Hong Kong: Hong Kong University Press, 2009, pp. 3–16.
38. H. J. Ladegaard, "Politeness, power and control: The use of humour in cross-cultural telecommunications," in *Professional Communication: Collaboration Between Academics and Practitioners,* W. Cheng and K. C. C. Kong, Eds. Hong Kong: Hong Kong University Press, 2009, pp. 191–209.
39. E. O. Wittman, and K. Windon, "Twisted tongues, tied hands: Translation studies and the English major," *College English,* vol. 72, no. 5, pp. 449–469, 2010.

2

Cultural Contexts in Document Design

Yiqin Wang

Harbin Institute of Technology

Dan Wang

Harbin Institute of Technology

To communicate effectively in international settings, it is often important to consider possible target languages and cultural contexts when designing technical documents. Cultural differences related to high- and low-context communication, aspects of language, and thought patterns can influence document design in terms of a number of factors including structure, content organization, amount of information provided, terminology, and the relationship between text and graphics. This chapter provides technical and professional communicators with guidelines for designing documents for effective intercultural communication. The approaches presented in this chapter can help readers from different cultural backgrounds optimally interpret the different kinds of information presented in technical documents.

2.1 The Challenges of International Communication

A major challenge to effective international communication is the need to adapt messages to the communication expectations of different cultures. In technical and professional communication, information is usually written in the context of one particular culture (i.e., that of the author). The underlying notion is that this information will later be translated or localized in order to share ideas with readers from other cultures. In many cases, cultural expectations and practices can affect how individuals from different cultures present or interpret spoken or written information. If not recognized and addressed, such differences can lead to miscommunication and confusion when materials designed for use by the members of one culture are introduced into the context of another. Such cultural expectations, moreover, can relate to everything from the language used to convey ideas to the layout and design of the documents—or other materials—that contain text. For these reasons, it is important to consider possible target languages (i.e., the language into which one is translating a document) and cultural contexts when designing technical documents. Accordingly, an effective understanding of cultural factors relating to document design can be essential to communicating successfully with audiences from other cultures.

While several factors need to be considered when developing materials for audiences from other cultures, areas that merit particular consideration include

- The overall structure of a document and of the parts of a document
- The organization of the content presented in a document
- The amount of information provided in a document
- The terminology used within a document
- The relationship between the text and the graphics that appear in a document.

Further complicating this situation are aspects of culture that can influence the interpretations of document design. These factors include high- and low-context communication (covered herein, in brief), language, culturally specific rhetorical strategies, thought patterns, and cultural differences in processing graphics.

The purpose of this chapter is to provide technical and professional communicators with a framework and guidelines for working in intercultural contexts. Such a framework and guidelines could, in turn, help technical and professional communicators identify different cultural expectations and preferences related to technical documents. These guidelines can also help individuals creating materials for audiences in other cultures to design effective technical and professional documents in terms of structure, content, organization, and graphics selection. By following such approaches, technical and professional communicators can help readers from different cultural backgrounds optimally interpret the information presented in a given document.

The specific objectives of the chapter are to examine aspects of culture and document design expectations in terms of

- **Developing** approaches for identifying cultural expectations relating to the design of different kinds of technical documents
- **Identifying** major cultural dimensions that can affect document design in international contexts

- **Explaining** how cultural dimensions influence the presentation of technical information in relation to the structure, content, organization, and use of text and graphics in a particular document
- **Discussing** how cultural dimensions affect the presentation of technical information in documents designed for individuals from other cultures.

To achieve these objectives, we, the authors, first examine the effects culture can have on document design. Next, we present a study of select automotive service literature from China and Germany and use this study to demonstrate how one can identify cultural differences in the presentation and perception of technical information. The authors also discuss the influence culture can have on communication expectations and note cultural differences between the way individuals in China and individuals in Germany react to and use different aspects of document design. In examining these topics, we present an overview of how a document analysis (i.e., reviewing a document in order to identify and analyze certain design features) can serve as a mechanism that technical and professional communicators can use to identify cultural factors affecting various aspects of document design.

2.2 Literature Review

Culture can be defined as the collective programming of the mind that distinguishes the members of one group from another in terms of norms, values, and attitudes [1]. The most significant cultural differences in technical documentation include those items that involve perception and thought patterns as they relate to the processing of graphics. These differences have an apparent influence on the presentation of technical information in relation to content organization and visual communication [2–5].

2.2.1 Cultural Differences in Perception and Thought Patterns

The most important step toward successful intercultural communication involves the appropriate perception of the intended subject matter by the intended audience. However, as Maletzke [6] notes, people from different cultures perceive the world differently. More specifically, people from one culture might pay attention to particular objects that might be noticed only slightly, or be completely overlooked or ignored, by individuals from another culture.

Cultural differences influence not only perception, but also ways of thinking. Linguistic relativity indicates that cognitive processes such as thought and experience may be influenced by the categories and patterns of the language that a person speaks and how they understand the world and behave in it. In other words, the nature of a particular language influences the habitual thoughts of its speakers. Different patterns of language yield different patterns of thought. The implications of this theory even prompted German educator Wilhelm von Humboldt to propose that thought and language were inseparable from each other [7].

In examining such relationships, Kaplan [8] suggests that the thought pattern that speakers and readers of English expect to encounter as an integral part of their communication is a dominantly linear sequence. That is, an initial premise is stated first followed by supporting evidence, and a conclusion is presented last. In Arabic, in contrast, thoughts are expressed as a complex series of parallel constructions. Kaplan also notes that, in some oriental languages, the development of thoughts may be said to turn in a spiral. In other words, the development of the paragraph tends to be turning in a widening gyre which moves around the subject and shows it from a variety of tangential views rather than looking at the subject directly. "Things are developed in terms of what they are not, rather than in terms of what they are" [8].

In an attempt to promote intercultural understanding in technical communication, Ding [4] studied the *Yi Jing* (*The Book of Changes*), which he regards as the first Chinese work of technical communication. In his examination, Ding reviewed both the kind and amount of information contained in the *Yi Jing* as well as how that information was organized within the text. He also performed a similar form of analysis on a modern Chinese instruction manual. As a result of this analysis, Ding claims that the *Yi Jing* advocates unity between a context and the individual events or objects existing in that context rather than viewing them as separate and unrelated entities.

According to this tradition, Chinese instructional manuals usually focus on contextual information instead of action-oriented instructions for task performance. Take the instructional manuals for installing a household water heater as an example. The steps in the US version of such a manual are clearly task-oriented actions that guide readers to perform specific tasks. In contrast, the steps in the Chinese manual focus on ensuring an ideal result of installing the water heater rather than guiding users to perform specific tasks [4].

To investigate cultural differences in visual communication, Wang [5] compared graphics from popular Chinese science magazines and instruction manuals with their counterparts from US sources. The results of this comparison indicate that, when presenting new scientific concepts, Chinese visuals tend to provide readers with more contextual information. The US visuals, on the other hand, are more closely integrated with corresponding verbal explanations. In a US manual on how to assemble a fan, for example, each action is explained in words and is illustrated via complementing pictures that appear in the same area of the text. The visuals that appear in parallel Chinese manuals, however, are not as closely connected to the related text to which they correspond.

Similarly, Maitra and Goswami [9] examined the visual design of a Japanese company's annual report and found very distinctive design features. They explain that, for the Japanese, the design of documents (e.g., page layout, pictures used, and formatting) is intended to impress the reader. For this reason, these documents contain design features such as aesthetics and ambiguity. In this case, the key objective is to present the company's image with a visually attractive document so that the reader will form a favorable impression of the related company. Japanese designers, therefore, often prefer to use as many pictures as possible—sometimes even images not directly connected to the company—in order to achieve the goal of making the document more beautiful/aesthetically pleasing. The rationale is that if these visuals can impress a reader, that person would then be more likely to read the actual text of a given document and thus gather the necessary information contained in that document. Contrary to the Japanese idea of maintaining aesthetics and ambiguity in the use of visuals in intercultural documents, the US concept of internationalizing graphics aims at clarity and cultural neutrality.

Other research also focuses on cultural diversity and its impact on communication. In so doing, such work supports the view that visual communication is shaped by culture. Tzeng and Trung's study [10], for example, indicates that the same visual symbols presented to observers from different cultures might generate contradictory responses. To examine this factor, Tzeng and Trung selected 10 graphics that were maximally representative of geometrical elements in the human-made and natural environments. These graphics were comprised of combinations of straight lines, circles, and waved or zig-zagged lines, and such features appeared in different degrees of complexity across examples. One-hundred college students from five countries were selected to participate in this study [10] based upon the idea that the graphics used in the study had maximum relevance in the cultures of Japan, the United States, Spain, Mexico, and Columbia. The results of the study indicate the existence of cultural differences related to the perceptions of visual factors and indicated that cultures seem to favor graphics and icons that are similar or indigenous to their national traditions.

2.2.2 Cultural Differences in Text Organization

Researchers have also studied cultural differences in the organization of a text [2, 3, 11]. Some studies, for example, examined United States and Japanese readers' comprehension of and preference for expository text that contains a thesis and that is organized either inductively or deductively. (Inductively organized texts describe supporting ideas first and then end with a general statement whereas deductively organized texts move from general information to specific information.) The results of these studies reveal that Japanese readers seem to better comprehend content organized inductively than content organized deductively in certain situations (e.g., when a passage presents the writer's view of the content in the form of a thesis or conclusion). Such prior work also indicates that Japanese readers seem to recall more information from inductively organized text than do US readers [2, 3].

By examining a variety of documents from China and the United States, Barnum and Li [11] demonstrate that differences in thinking patterns affect both the structure and organization of a document and the arrangement of information within a document. They point out that US technical documents tend to reflect the Western preference for an analytical thought pattern. US writers, for example, tend to frame an argument and then divide the presentation into chunks of information that supports the argument. Within this context, headings and subheadings provide readers with a road map of the various points covered in the overall discussion. Chinese writers, in contrast, tend to employ synthetic, or integral, thinking patterns as the basis for organizational structure. The cultural basis for this preference has been characterized as "the relational style" (i.e., more holistic and context-based) versus the "analytical style" (i.e., guided by universal rules) dominant in North American culture.

Chinese thought strives for unity between events or objects and their given signs or symbols. To understand how the relational style affects the Chinese view of the world, we can use traditional Chinese medicine, which views the exploration of illness in a markedly different way than conventional Western medical practices. When treating a patient, for example, a doctor of traditional Chinese medicine looks at the patient as an organic whole, viewing the various parts of the body as closely interrelated and related to the external environment rather than as individual units that merit focused, individualized examination (versus being viewed as part of the overall body).

In addition to the literature reviewed here, other researchers in intercultural technical communication investigate multiple themes by conducting cross-cultural comparisons. Mirsahfiei's [12] observations of and research on Arab, Afghani, Iranian, Indian, and Pakistani students found that cultural factors affect important areas of technical communication including content, style, structure, and format. These factors also seem to affect the role of the student in the learning process, the student's thinking process, and the student's attitude toward and treatment of information. The results of Mirshafiei's study thus seem to indicate there is a need for more research on the problem of cultural influence in technical communication. Additionally, Mirshafiei thought that these findings provided a foundation others could use both for further research on how culture influences technical communication and to design and teach technical writing courses to foreign students and to professionals from other cultures more effectively.

Thrush [13] echoes these ideas when she points out that research in anthropology, cognitive psychology, linguistics, and writing theory has identified several factors that vary within languages and cultures and affect the way readers read and interpret texts. Those factors include world experience, amount of common knowledge shared within a culture, and the hierarchical structure of society and the workplace. They also encompass culturally specific rhetorical strategies for presenting information and ideas as well as cultural differences related to how graphics are processed by individuals.

2.2.3 Creating a Context for Understanding

Previous research in intercultural technical communication mainly compares aspects such as vocabulary, visuals, and text across multiple cultures. Certain studies focus on one specific theme, such as visualization and text [5, 10]. Other research investigates multiple themes by conducting cross-cultural comparisons [12, 13]. In other instances, researchers have examined how two cultures compare in relation to a particular technical communication topic such as visualization or the organization of a text (see References 2, 3, 5, and 10) while still other investigators have studied multiple aspects of technical communication as it is practiced in one culture [4, 9].

Despite this work, there has been relatively little research that seeks to compare technical communication practices between two cultures in terms of multiple design factors such as the combined aspects of the overall structure of documents, the organization of text in documents, and the use graphics in different documents. This chapter, in contrast, presents a comprehensive framework that identifies the influence of cultural contexts on the design of technical documents. This framework can, in turn, help technical and professional communicators better understand cultural differences and cultural factors that may affect the design expectations and practices of various cultures.

In presenting this framework, the authors compare technical communication in German and Chinese communication contexts. In so doing, they examine materials across multiple dimensions. The objective of this comparative approach is to examine cultural differences related to how groups present and perceive technical information. To this

STUDY DESIGN 25

end, the authors use examples of technical communication from Germany and China to address these questions:

- How do these two cultures differ in the presentation and perception of technical information?
- What kind of cultural factors may be responsible for these differences?
- How do the cultural factors affect the design of technical documents?

Through such an examination, the authors explore cultural contexts in technical document design and connect them to real-world examples via the analysis of documents as well as through a user study involving the two cultures.

2.3 Study Design

Cultural differences usually leave some imprint on certain aspects of technical documents. Such imprints, moreover, might contain different features and emphases depending on the culture that created a given document. A document analysis can therefore serve as a valuable mechanism to identify cultural differences related to how technical information is presented.

Perceptions of technical documents are similarly influenced by cultural background. Thus, people from different cultures might have different perceptions of the same technical documentation. And, in work environments, individuals from different cultures may have different perceptions of the information being conveyed in a technical document. A user study can thus help identify these differences.

To address such factors, the research reported in this chapter employs both the process of document analysis and user studies to examine how documents are created by and perceived by different cultural groups. In particular, the following approach was used to examine such items:

- **Document Analysis**—In this analysis, the structuring, content organization, terminology and illustrations (graphics) used in technical documents created by German and Chinese individuals for members of those same cultures were studied in order to identify cultural differences in the characteristics of technical information and find out how they exert their impact.
- **User Study**—To minimize the bias resulting from handling different products or documents, our user study was conducted in three Chinese and two German automobile workshops dealing in the same brand of automobile and handling the same products and technical documents. The workshops were studied to obtain background information that would help us identify cultural differences. The participants were mechanics who worked in the workshops. Seventeen mechanics in three Chinese workshops and 16 mechanics in two German workshops participated in the study. They had similar jobs related to automobile maintenance and repair.

This two-part approach thus allows the authors to examine the complexities of different cultural aspects that can affect the creation and the use of technical documentation.

2.4 High- or Low-Context Culture and the Extent of Explicit Description

2.4.1 High- or Low-Context Culture

High-context (HC) communication implies that most of the information exists in the shared knowledge of the communicators, while very little is in the coded, explicit, transmitted part of the message. As a result, very little information needs to be conveyed explicitly among individuals, for all involved are assumed to be able to intuit meaning from a given context. Low-context (LC) communication is just the opposite; that is, most of the information is vested in the explicit code and information is stated directly rather than assumed to be conveyed via a given setting [14].

This distinction means members of low-context cultures often state information directly and explicitly while individuals from high-context cultures tend to favor more indirect methods for conveying information. For example, a person from a low-context culture would directly state "Please turn up the temperature in the room," whereas someone from a high-context culture might use the more indirect "Do you find it cold in here?" to convey the same message. Similarly, writers in high-context cultures do not need to provide a great deal of background information or to spell things out precisely when discussing a topic [13]. Rather, it is assumed that the audience with whom one is interacting has the background knowledge needed to correctly intuit the objective of a message. In terms of technical documentation, attention should be given to the features of high- and low-context cultures. To be more specific, technical documents in a high-context culture do not need to be as explicitly written as those in a low-context culture. For example, documents do not need to be as formally structured for readers in a high-context culture as for those in a low-context one.

2.4.2 A Study of Chinese and German Automobile Literature

It is generally acknowledged that Northern and Western (Germanic) countries are more low-context, explicit cultures that use direct, linear discourse in communication. By contrast, Southern (Latin) and Eastern (Asian) cultures tend to be identified as more high-context, implicit cultures that prefer indirect, digressive communication patterns [15]. Such factors mean that German textbooks and service manuals tend to be more finely detailed and structured than Chinese ones. Additionally, such factors could mean that the overall structure of documents does not have to be as carefully crafted for a high-context culture as it does for a low-context one [16]. Such differences became important in the document analysis the authors undertook to examine materials created by members of these two cultures.

2.4.2.1 Document Analysis—Structuring There are basically three general structuring methods, or strategies, for writing technical documents: product-oriented, task-oriented, and user-oriented.

- In a product-oriented structuring, the writer focuses on either the product or on its components. Such a structure often resembles menus which list the components and assembly group of a product, describe the purpose of these components, and note the related function and process of the component parts and the overall product.
- In a task-oriented structuring, the writer explicitly explains the tasks to be completed by using the product. In such cases, the text is generally organized according to the possible task, employment, or operation that a user should complete by using a given product.
- In a user-oriented structuring, the chronology of operations activities is taken as the main thread in the structuring of technical documents [17]. Such materials might look like a list of activities to be performed in a chronological order.

The documents the authors examined in order to study structuring and culture consisted of

- A German textbook used to train automotive mechanics [18]
- A Chinese textbook designed to achieve the same objective with Chinese automotive mechanics [5]
- A German Service Manual for autotype series 129 of Mercedes-Benz
- A Chinese Service Manual for autotype CA770 of First Automobile Works (FAW)

These texts were selected for analysis because they were similar in terms of their overall informational intent, contents, and indented audiences. In other words, both sets of manuals were designed and used for the education and training of automotive mechanics.

2.4.2.2 User Study—Theoretical Troubleshooting Test In the theoretical troubleshooting test conducted by the authors, the participating mechanics were provided with documents they were to use in troubleshooting the stated malfunction: "The engine does not start while the starter is cranking up." The documents provided to the mechanics contained information on the following topics:

- The structure of the engine fuel supply
- The task of the engine fuel supply
- The function of the engine fuel supply
- An overview of the automobile's electronic systems (CAN—Date Bus Schema)

- An arrangement of parts
- An overall functional schema; and so forth

In this situation, the mechanics had to

- Read and review the documents in order to find the instructions corresponding to the task they wished to undertake
- Understand the information related to performing that task
- Apply the related information in a way that allowed them to perform the desired task

This kind of activity constitutes a complex mental process that includes:

- Goal setting, in which the goal of the task is formed by the user
- Category selection, in which the major categories of information within the document are identified
- Information extraction, in which relevant details within a selected category are thoroughly identified
- Cognitive integration in which extracted information is combined with previously obtained information [19]

In the user study, the authors used a combination of observation and interviews to collect data. For this process, the participating mechanics were asked to

- Select the informational themes they needed to review in order to accomplish the desired objective (i.e., identifying and correcting the problem)
- Identify the importance of informational themes (high, medium, or low)
- Assign priorities to each theme accordingly.

2.4.3 Findings Observed

The overall structure of the textbooks was oriented toward products and their components (i.e., product-oriented structuring) in both cultures. The chapters within the books, however, were structured for users (user-oriented structuring) in China and for products (product-oriented structuring) in Germany. The structure in Chinese service manuals was based on the chronology of different operations activities users should carry out. These activities included maintenance, disassembling, checking, assembling, and adjustment for components such as an engine. The German textbooks and service manuals, in contrast, were more comprehensively structured with greater detail than the Chinese ones. For example, the catalog of the chapter "chassis" included only steering and braking in the Chinese textbook instead of listing all of the contents such as steering, suspension, vibration absorption, wheel suspension, and so on, as was done in the German textbook.

The results of the troubleshooting test revealed that the Chinese mechanics in the Chinese workshops considered three to four informational themes as very important, three themes to be of average importance, and three themes to be considered unimportant. The German mechanics in the German workshops, in comparison, considered five themes as very important, and four to five themes to be of average importance. Because the Chinese and German workshops deal in the same brand of automobile and the mechanics had similar jobs related to automobile maintenance and repair and had the same products and technical documents, the user study would help us identify cultural differences. These results suggest that the Chinese mechanics who participated in this study needed less information than their German colleagues when it came to troubleshooting.

These observations could be connected to the high-context culture of China and the low-context culture of Germany. In a high-context culture, people might need little coded, explicit information to be stated (either in speech or in documentation) because individuals might be well-informed about the environment in which an interaction takes place (e.g., in which a document is read and used). As a result, most of the information exists in the shared knowledge of the communicators (in the case of the troubleshooting documents studied here, the author and the reader of a given document). In contrast, in a low-context culture, individuals might prefer to be provided with more information that would be explicitly coded and to have the context explained in detail (i.e., all items stated directly within the context of the document, for the author cannot assume what the reader already knows about the topic covered in a given document).

2.5 Thought Pattern and Content Organization

2.5.1 Thought Pattern

The idea that culture profoundly influences the contents of thought through shared knowledge structures has been a central theme in modern cognitive anthropology [20]. Scholars such as Nisbett and Peng [21] have, in turn, examined such relationships by exploring the idea that East Asians and Westerners have different ways of reasoning, namely, holistic (East Asian) versus analytic (Western) reasoning. Within this context,

- Holistic thought involves an orientation to the context or field as a whole, including attention to relationships between a focal object and the field. Holistic thought also generally includes a preference for explaining and predicting events on the basis of such relationships.
- Analytic thought, in contrast, involves detachment of the object from its context and a tendency to focus on attributes of the object to assign it to categories. Analytical thought also tends to include a preference for using rules about the categories to explain and predict the object's behavior.

So, a holistic thinker (such as someone who is culturally Chinese) might hold the "view that the world is a collection of overlapping and interpenetrating stuffs or substances" whereas an analytical thinker (someone who is culturally German) might be inclined to "see the world as a collection of discrete objects which could be categorized by reference to some subset of universal properties that characterized the object" [21].[1]

2.5.2 A Study of Chinese and German Automobile Literature

The thought patterns, regardless of whether they involve synthesis (holistic thought) or analysis, influence the organization of contents in documentation. If the synthesis thought pattern predominates in one culture, the content about a system or a component tends to be organized in the technical document with interrelations and context. Hence, a document designed according to a *synthesis* (holistic) thought pattern—such as that often associated with Chinese culture—might begin by providing general information and organize the contents of the overall document as if that item was one unified entity with interrelation among its elements. If the *analysis* (analytic) thought pattern—often associated with German culture—is predominant in a given culture, its technical documents would tend to be structured with specific and separate elements (e.g., be divided into heading-based sections and sub-sections) [16]. Such a document might start from concrete subsystems and be organized with specific and separate elements.

2.5.2.1 Document Analysis—Content Analysis Many technological systems are integrated into automobiles, and this factor makes the presentation of information and instructions associated with such technologies relatively complex. The organization of contents about such complex systems also plays an important role in the users' comprehension of the systems. The analysis of such content organization in technical documents can help researchers identify how cultures exert their preferred cognitive approaches and systems on technical communication.

For this study, the authors of this study examined the organization of a textbook chapter on the topic of an automobile engine's fuel injection system, and did so for chapters that appeared in parallel German and Chinese textbooks [18, 22]. (As noted earlier, both books were designed for the education and training of automotive mechanics in the respective cultures.) The descriptive text found in these two examples includes information about the product structure, the features, and the combination and function of fuel injection systems. Specifically, the analysis the authors did of these two texts included an examination of the organization and interrelation of the content about the system including the treatment of topics such as

- Type of injection system (KE-Jetronic; HF- and P-Motronic)
- Type and objective of the control

---

[1] Evidence for the distinction between holistic and analytic thought comes from the study of ancient Chinese and Greek philosophy, mathematics, and science [35]. To the ancient Chinese, matter was continuous and interpenetrating, and events were the result of an interaction between the objects and the field. Thus, neither formal categories nor formal logic played much of a role in mathematics or science. Though the ancient Greeks were far behind the ancient Chinese in terms of technological achievements, it was the Greeks who invented many of the concepts and perceptions that have come to characterize science, defined as the explicit causal modeling of events based on a formal system of rules and categories. The ancient Greeks did not understand that all action is a result of the interaction of an object in a field of forces, but they did have the beliefs that focused on the object at the expense of the context.

- Construction and function of the sub-system (air-, fuel-, ignition-, and control-system)
- Components of the system (sensors, control unit, actuators, and fuel facilities)
- Its objective, construction, and function

This type of text has the function of explicating concept knowledge for users or presenting context for instructional text [17].

The methods of organizing the content in the two different books were analyzed to find out whether differences in presenting and organizing information existed as well as to determine how the content of the two parallel texts might be different. In essence, the focus of this document analysis was to determine how the system was described in each text (i.e., was it described by providing the general information first and then followed by interrelation among its elements as one entity—a more holistic approach, or was it described starting from concrete subsystems and organized with specific and separate elements—a more analytic approach).

The analysis of the descriptions of the engine fuel injection system provided in the two textbooks indicates that there are differences in the organization of content in the parallel Chinese and German documents. In the Chinese textbook, a system is described starting from the general information (e.g., types and objectives of the control, concept and development, types of injection system, structure and function of the system). The German textbook, by contrast, starts from the concrete system, such as the K-, KE-Jetronic, L-, LH-, and P-Motronic system. In the Chinese textbook, a system was structured on the basis of an entire interrelation or context in the documents (e.g., general information on types of injector systems). In the German textbooks, however, a system is presented independently and separately structured as individual elements (e.g., the KE-Jetronic). Such differences indicate that

- Germans tend to engage in context-independent and analytic perceptual processes by focusing on a salient object independently of its context.
- Chinese tend to engage in context-dependent and holistic perceptual processes by attending to the relationship between the object and the context in which the object is located.

Such factors have interesting implications for individuals who might wish to try to design more "universal" technical documents for users/readers from both cultures.

2.5.2.2 User Study—Theoretical Troubleshooting Test

In the theoretical troubleshooting test mentioned previously, the mechanics were provided with documents for troubleshooting. They were then asked to review these documents and select the informational themes and assign priorities/rank these themes accordingly. The Chinese mechanics had more or less identical orientations in troubleshooting. More specifically, the information search sequence of the majority of the Chinese mechanics was from general information (e.g., the overall function schema and overview of the electronic system of an automobile) to concrete information (e.g., sensors and actuators, as well as arrangement of parts). In contrast, half of the German mechanics in the test had an information search sequence that moved from concrete to general: that is, from the arrangement of parts to the overview of the electronic system of an automobile.

2.5.3 Findings Observed

Based on the results of the comparative study of Chinese and German automobile literature described in this chapter, the authors have identified the following trends relating to culture and communicating technical information:

- In the presentation of the technical information, the analysis of content organization revealed that systems were structured as entities with interrelations or context in Chinese documents, and specifically and separately structured as elements in their German equivalents.
- The troubleshooting test showed that Chinese mechanics favored starting from the general information; by contrast, the German mechanics preferred starting from concrete subsystems in perceptions of the technical information.

2.5.3.1 Findings on Patterns of Thought These observations could be explained by differences in thought patterns (i.e., the distinction between synthesis or analysis thought patterns). The Chinese, with synthesis thought patterns, tend to join separate elements in their way of thinking; whereas the Westerners (i.e., the Germans), with analysis thought patterns, usually follow an imaginary dissection of a whole into its parts and of a system into its elements [23].

2.5.3.2 Connections Between Thought Patterns, Language, and Writing The authors wish to note that the Chinese language seems to parallel the Chinese thought pattern noted here in terms of emphasis on the general state of being, on the natural being, and on the context of things. Chinese thought patterns, which are usually described as holistic and relational [5], have reflections on its language.

Within this context, word formation offers an interesting example for studying such connections. Chinese word formation mostly consists of suffixes (*–qi* (apparatus,), *-ji* (machine) and *–xitong* (system)) used to indicate that many parts combine and interact with one another and work as a whole [16]. Such relationships are further reflected in Chinese medicine, which assumes the body is a unity in which the elements are organically combined with each another. It adapts itself for the totality and the whole [24, 25].

Likewise, the way dates and addresses are formatted parallels the ideas of differing thought patterns seen in the Chinese and German cultures. That is, in Chinese culture, the sequence of date is displayed as year-month-date (i.e., moving from the greater unit or whole of the year to the more focused date), and the address is usually written in the sequences of country, province, city, street, the number of the street, and name of the receiver in China (again, moving from the greater/larger unit to the more focused one). In Germany, by contrast, dates are written just the opposite as day-month-year (i.e., specific/smaller unit to general/larger unit), and the address is written in the order of the name of the receiver, number of the street, city, province, and country (again, moving from the specific, smaller unit to the larger, more encompassing one).

2.6 Cultural Contexts in Text–Graphic Relationships

The objective of this section is to discuss how the text–graphic relationship differs in technical documents between the Chinese and German cultures. The possible reasons for such differences are also examined in terms of cultural factors, and special attention is paid to cultural aspects of thinking and learning as well. Through this examination, the authors explore the influence of culture on the characteristics of technical illustration in relation to the text–graphic relationship. (They do so by both the analysis of documents and a user study involving participants from the two cultures.)

2.6.1 Text–Graphic Relationships

Normally, graphics are rarely presented alone (i.e., without corresponding text descriptions) in technical documents. Rather, written descriptions and graphical information are usually connected within technical documents in order to better address the interrelated cognitive process of text and graphic comprehension [26]. For this reason, when used in technical documents, graphics are often arranged next to the corresponding text on the page in order to facilitate reader understanding.

In general, graphics repeat the information presented in the text and clarify what might be vague or too abstract to grasp [26]. That is, they complement verbal presentation of the content, making it more accessible to the reader. For these reasons, there may be various ways to group graphics and relate them with the corresponding text.

To help us better understand such relationships, Ballstaedt [26] defines two different types of text–graphic relationships: elaborative and redundant. The *elaborative* relationship is one when the text and graphics are in perfect complement. In a *redundant* relationship, a detailed text description is provided when the graphics are sufficiently illustrative. Within this context, information about a product's structure, features, combination, and function can be illustrated either briefly or complexly. The accompanying text can thus be implicitly or explicitly described. Consequently, graphics can be described on the basis of an entire context or separately described as equivalent individual elements.

2.6.2 Cultural Factors and Their Impacts on Text–Graphic Relationships

Visual communication and the relationship between text and graphics may be shaped by culture. This view is supported by research focusing on cultural diversity and its impact on communication. In essence, visual design must reflect social and cultural values because visual communication is closely bound to experiences [27]. As a result of such factors, the same visual symbols presented to observers from different cultures may generate quite contradictory responses [10]. Chinese visuals, for example, tend to provide more contextual information to readers. German visuals, on the other hand, are more closely integrated with corresponding verbal explanations [16, 28].

Visual comprehension ability is believed to be affected by the language environment in which individuals grow up, be it a culture that represents language via pictographs (such as Chinese) or phonetic representation (such as German). If this visual comprehension ability (i.e., the ability to draw meaning from visuals) is strong in one culture, the technical documents designed for that culture should be presented in an elaborative way with more graphics and less text. (Such cultures would be characterized by synthesis thought patterns and would generally be considered more high context.) So, in such cultures, the illustration of a system or a component should be described with interrelation and context as a whole. In a culture with analysis thought patterns (generally a low-context culture) by contrast, illustrations should generally be explicitly described via specific and separate elements [28]. These ideas were used as a foundation for a case study that combined document analysis and user testing to examine such relationships.

2.6.3 A Study of Chinese and German Automobile Literature

Previous research on this topic indicates that culture exerts influence on characteristics of visual communication [5, 9, 10]. To further examine such ideas, the authors decided to conduct a document analysis, for such an approach can facilitate the identification of cultural differences in the visual presentation of technical documents. The authors also conducted a parallel user study of related documentation in order to identify cultural differences relating to the perception of illustrations used in similar kinds of technical documents.

In the document analysis phase of this project, the authors studied the graphics in automotive textbooks and service manuals according to the arrangement of text-graphics and text–graphic relationships in content. For this analysis, the authors used parallel texts (i.e., a description of a fuel engine injection system) in one German [18] and one Chinese textbook [29]. Both textbooks, with regard to their overall informational intent, their content, and their readership, reveal that they had a similar audience; they were designed and used for the education and training of automotive mechanics. The description texts about the fuel engine injection system were chosen as corpus. This text included information about the product's structure, features, and functions in relation to the other parts of the system and its parts. The related text also contained many illustrations to explain the construction and function of the system.

In this study, the authors analyzed how the language description and graphic information were connected and interrelated. Instructional texts about the disassembly of a transmission case in the Chinese and German handbooks [30, 31] were chosen as the corpus, which are designed for the same target group—automotive mechanics to explain how to disassemble the transmission case, to analyze how often the illustrations are used and how the text and graphics are arranged. The instructional texts are designed to provide automotive mechanics with information about the process of the transmission case disassembly. The results of this document analysis are summarized here.

2.6.3.1 Findings of the Document Analysis Phase Research As far as the text–graphic relationship in content was concerned, the authors noted the following trends:

- More graphics and a more elaborate relationship were used to convey information in the Chinese texts, while fewer graphics and redundant relationships were used to convey ideas in parallel German texts. In repair instructions on the transmission case disassembly, for example, there were 25 graphics in the Chinese manual, but only 12 graphics in the corresponding German manual. Both manuals, however, presented the same content relating to what mechanics should do to disassemble a transmission case.
- The Chinese textbook showed a strong complementariness between the text and graphics with the text usually providing additional information not conveyed in the graphics. For instance, in a Chinese textbook, the text that accompanied the illustration of the system L-Jetronic offers information that is not covered in the graphics: "The air flow in the petrol injection system is measured in volume, that is, the analog signal is transformed into digital signal by measurement of the air flow which is absorbed into the cylinder, and the processor calculates corresponding fuel flow to control the optimal air/fuel ratio" [29].

In contrast, the German textbook displayed a parallelism between the graphics and the text in that the text offered the same kind of information—be it in more detail—conveyed in the graphics. In the case of the illustration of the L–Jetronic System, the corresponding text in the German version described the whole system as follows:
- *The fuel system (fuel -container, -pump, -filter, pressure regular, distribution liner)*
- *Sensors (air-flow meter, engine temperature sensor, constant time switch, Lamda-Sonda, throttle-valve switch)*
- *Actuators (injection valve, cold starter valve, accessory valve) [18].*

The comparison of the texts reveals most of the illustrations about the system structure, combination, and function that appear in the Chinese textbook are brief. However, the parallel example from the German textbook is relatively complex by comparison. In the Chinese textbook, the accompanying text tends to provide more contextual information on the basis of an entire interrelation. In the illustration of the fuel supply system, for example, the illustrated system is described as a whole with interrelation: "Fuel is supplied from the fuel tank by the fuel pump, and then is filtered by the fuel filter to eliminate impurity and water, after that the fuel is delivered to the absorber in order to reduce the fluctuation. In this way the fuel with definite pressure will be delivered to the chief pipeline, then through the branch pipeline to the injector in pro cylinder. The injector drives the nozzle and sprays an adaptive amount of fuel ahead of the inlet valve, then the air-fuel mixture will be taken into the cylinder" [29].

In contrast, the German textbook displays a more explicit description of the illustration. Additionally, the individual elements of the illustration are described separately. The corresponding text accompanying the related illustration describes the whole system as follows:

"18. Fuel Supply Equipment
- 18.1 Parts of the fuel supply equipment The fuel supply equipment has the function for inject equipment to supply enough amount of fuel which is free of blow holes, with needed pressure and without pressure fluctuation.
  - 18.1.1 Fuel tank
  - 18.1.2 Facilities for ventilation ...
  - 18.1.3 Fuel reserve indicator ...
  - 18.1.4 Fuel pipe ...
  - 18.1.5 Fuel filter ...
- 18.2 Fuel supply pump ...
  - 18.2.1 Function of the fuel supply pump ...
  - 18.2.2 Electric fuel supply pumps ..." [18]

Each section was described in detail, which included a description of composing parts, the features of the parts and the working mechanism, and so on.

It should be noted that both textbooks had similar audience and were designed for and used in the education and training of automotive mechanics and the writing and presentation style of the textbooks were mainly influenced by the culture. There may exist some limitations in this study of comparing German and Chinese manuals, such as a style difference between companies and their in-house formats, or a difference in the writers' style for these manuals.

2.6.3.2 Findings of the User Study Phase Research *Comprehensibility* of a document refers to the readers' perception of the document's readability, understandability, organizational coherence, and the accessibility of textual meaning. *Understandability* can be measured by the readers' perception of having grasped the purported meaning of the text and its purpose [31]. A functional description of a new electronic system for diesel engines, called the common rail injection (CDI) system, was chosen as the text to be examined in this phase of the research process.

The CDI system is complex and the documents describing how the system functions are presented with a combination of text and graphics. The documents were reduced by the authors of this study and represented with more graphics and less text. That is, the text description which is illustrative in the graphics was abridged. The chosen documents, for example, include

CDI generalization
- Short introduction of the CDI system and its advantages
- Construction and function of the system

CULTURAL CONTEXTS IN TEXT-GRAPHIC RELATIONSHIPS

- Control unit CDI—generalization
- Input–Output signals with a diagram
- Arrangement of the control unit CDI (only one diagram) [32]

The same groups of mechanics as mentioned earlier were asked to read the documents provided by the authors and then answer the following questions which were designed by one author who had educational experience in automobile electronic system and knew the system and the answers well:

- What are the tasks of the rail?
- Which parts will be needed for high-pressure regulation?
- How is a softer combustion process achieved?
- Which two factors regulate the dozens of injector?
- What is the use of identification of spark advanced angle of Cylinder 1?
- Which parts supply the control unit CDI (N39) with voltage?
- How does the control unit CDI (N3/9) identify spark advanced angle of Cylinder 1?

The questions could only be answered correctly if the mechanics had grasped the meaning of the texts and understood how the CDI system worked. Because the documents chosen were presented with more illustrations and less text, it was possible to determine the differences in comprehensibility of visualization and text as well as culturally relevant visual comprehension.

The results of the user study indicate that differences exist between the Chinese culture and German culture with respect to the comprehension of texts and graphics. The research done at this phase indicates Chinese mechanics, in general, have a better understanding of graphics than the German ones. During the comprehension test, for example, the mechanics were asked to answer questions about the tasks, construction, and functions of CDI System. Moreover, these individuals were asked to perform this task using only the documents that had been condensed in a way that had more graphics, but less text. It turned out that more questions were correctly answered by the Chinese mechanics (82%) than the Germans (63%), even though the latter were more familiar with and had more general knowledge about the system.

2.6.4 Findings Observed

Through this research and through a closer examination of the findings noted here, a series of trends or patterns involving culture and image use in technical documents began to emerge. The three overarching or more prevalent trends included the following:

- The comprehension test and graphic analysis indicates the Chinese mechanics had a better understanding of graphics than the Germans.

- The text–graphic relationship appears to be elaborative in the Chinese documents and redundant in the German ones. That is, in the Chinese textbooks examined, readers can find little written information that parallels what appears in the graphics. In the German textbook, however, readers can easily find the equivalent/parallel text descriptions of information that appears in graphics.
- The Chinese repair instructions contain more (almost twice as many) graphics as do the parallel German instructions.

With these findings and trends in mind, the question becomes: How are they connected to cultural factors and to the cultural backgrounds of individuals? The next section of this chapter presents ideas on the nature of such connections.

2.7 Cultural Backgrounds

The authors believe that the findings of the study can be explained in terms of cultural differences related to language, thought patterns, and high-/low-context differences.

2.7.1 Aspects of Language

The language environment (pictographs or phonetic language) in which individuals grow up affects their visual comprehension ability. Language environment, in turn, often connects to or affects the ways in which the brain operates in relation to processing information.

The human brain consists of a right and left hemisphere; the right hemisphere is mainly responsible for graphics and space, emotions, analogy, creativity, and instinct. The left hemisphere, by contrast, is mainly responsible for language, logic, digital information processing, data analysis, and recognizing objectives [17]. With these factors in mind, Jones [33] points out that the processing of phonological elements rests with the left hemisphere, while the processing of semantic visual elements rests with the right hemisphere. This factor is consistent with the idea that there are important hemisphere differences in the processing pictographic systems of writing such as Kana and Kanji (Chinese characters). As a result, both the right and left hemispheres of the brain are trained at the same time as individuals learn to read and write using Chinese characters. By contrast, only the left hemisphere tends to be trained when people learn phonetic characters (e.g., the letter-based alphabet and related writing system used in German) [34].

Chinese characters are generally pictographic; in other words, the meaning of the character is directly connected with its form. In learning to read and write using such characters, which fosters bi-hemispheric brain use, the connection between form and meaning is stronger than between phonetics (pronunciation) and meaning. The dominant trend in traditional Chinese thought reflects such factors as seen in a well-known motto from *The Book of Changes:* "The words are here [something immediate] while the meanings are there [something beyond]." Such a mode of thought has dominated the best minds in China for more than 2000 years [35].

The letter-based German system of writing, by contrast, is phonetic in nature. As a result, the reader (re)builds a word according to the letter that makes a particular sound (pronunciation). In learning to read and write in German, which fosters monohemispheric brain use, the meaning of the words is mainly connected with phonetics (pronunciation).

2.7.2 Thought Patterns

In the Chinese textbook analyzed by the authors, the text that accompanies images tends to provide more context information on the basis of an entire interrelation. The wording that appears in the German textbook reviewed by the authors, however, uses more explicit description of the illustration in terms of individual elements (e.g., sensors and actuators for the system "L-Jetronic") rather than as an overall whole. The German text also offers detailed descriptive information which is already partly covered in the related graphics.

This observed difference can be explained in terms of variations in thought patterns (i.e., synthesis or analysis thought patterns) between the two cultural groups. The Chinese, who use more of a synthesis thought pattern, tend to join separate elements together in their way of thinking. For instance, the text accompanying the illustration of the system "L-Jetronic" described how the parts (sensors, actuators) are interrelated and how the fuel system worked. By contrast, Westerners (e.g., the Germans), who tend to use the analysis thought pattern, usually follow an imaginary dissection of a whole into its parts, and a system into its elements [23].

2.7.3 High-Context versus Low-Context Communication

The authors believe that whether the corresponding text provides implicit or explicit information in textbooks is connected to cultural differences involving high- and low-context communication. The Chinese textbook analyzed here, for example, provides implicit and contextual information because the Chinese culture is more high-context. For readers from a high-context culture, it is not necessary to spell out every detail about real objects because they can comprehend most of the information existing in the shared knowledge of the user and in the context of the environment. For the more low-context German culture, however, such explicit displays of information—particularly in textual form—are expected, if not needed, in relation to understanding and using information.

2.8 Applying Ideas to Training in Technical and Professional Communication

Computer technology and the Internet are making intercultural communication increasingly faster, more frequent, and more visual in nature. To ensure the success of communication in this context, it is important for individuals working in professional communication to be aware of how the characteristics of the target culture can affect interactions—particularly interactions that rely on visual presentations of information.

Technical and professional communicators therefore need to understand how languages, patterns of thought, and even underlying philosophical differences can affect the presentation of technical information in different kinds of documents. Such factors are particularly acute in relation to how individuals perceive, interpret, and use information presented via illustrations. Through understanding such factors, technical and professional communicators can create visual elements—and visual displays of information—that can be accurately and effectively comprehended by individuals from other cultures.

2.8.1 Applying Ideas

The study presented in this chapter provides technical and professional communicators with a comparative framework for understanding the interrelationship of culture and expectations of visual communication in technical documents. As a result, the ideas discussed here can help technical and professional communicators understand and identify cultural differences related to the visual presentation of information in technical documents as well as how cultural factors can affect the perceptions of illustration used in such documents. Students in technical and professional communication can, in turn, use certain exercises to apply the ideas of culture, design, and professional communication examined here in order to better understand such factors and to be better prepared for the modern, global workplace. The exercise presented in the remainder of this section can be used by instructors to introduce students to such ideas. Instructors can then modify this foundational exercise to examine aspects of culture and visual design—particularly the interrelationship of visuals and images in technical documents—in a broader context and with different cultures.

2.8.2 An Overview of the Proposed Exercise

The objective of this exercise is to introduce students to an analytical framework they can use to identify cultural differences relating to the visual presentation of information in technical documents. Students can also use this framework to consider cultural differences when designing documents for individuals from a variety of cultures.

2.8.2.1 Selecting Materials for the Exercise Select a group of representative/common documents, that is, a corpus, preferably the same corpus in two languages, from technical documents in the industry in which students will likely find themselves working across cultures. In selecting materials to include in this corpus, instructors should try to address two central criteria:

- Select descriptive text (e.g., functional description for a system or an aggregate) and make sure that text includes information about the product structure, features, combination, and function. This factor is important, for this type of text serves to explicate concept knowledge to users or present context for the instructional text. As a result, such text usually contains many illustrations to help explain the construction and function of the system.

- Be sure that the materials chosen for the corpus come from two different cultures and that they are "equivalent." That is, the texts selected for inclusion in this corpus should address the same function, the same object, and the same target group. This parallelism is important because cultural differences in the visual presentation in technical documents can be identified only if the same system or aggregate is presented for the same function, object, and target group.

Once the instructor has selected the appropriate materials for such a corpus, students can use them to perform a review exercise designed to examine cultural differences related to the uses of visuals to convey technical information.

2.8.2.2 A Four-Step Process for Using the Exercise

To help students use the materials in the corpus to examine ideas of culture and visual design, the instructor should have students perform the following four-step process:

- **Step 1:** Analyze the materials provided by the instructor in order to determine how the language description and graphic information are connected and interrelated—whether they have an elaborative or redundant relationship. The objective of this step is to understand and grasp whether the text–graphic relationship is elaborative or redundant. The material should be translated into the language that most of the class have a good understanding of.
- **Step 2:** Compare the visual presentation of the corpus from the two cultures. The comparison relates to the text–graphic relationship in contents (elaborative or redundant relationship), and how often the illustrations are used. The differences in the visual presentation in relation to the text–graphic relationship should be identified in this step.
- **Step 3:** Analyze the cultural background of different audiences by applying the cultural factors as introduced in the previous parts of this chapter. This step tends to help students familiarize themselves with cultural factors such as high/low context, holistic/analysis thought patterns, and language, and to help students develop cultural sensitivity.
- **Step 4:** Identify the major dimensions of culture in the visual presentation, and discuss their effects on the text–graphics relationship. In other words, try to find if there is a more acceptable visual presentation in a particular culture.

Once this exercise is completed, the instructor should require students to use the results of this exercise to write a paper that examines or addresses the following points:

- **Title of the exercise:** Cultural context in the visual presentation—a study of… (e.g. "A Study of Chinese and German Automobile Literature"). The title of the paper should note which countries/cultures were compared and which industry was examined in doing this comparison.

- **Choose a corpus:** Explicate how and why the literature of the corpus used for analysis was chosen and what these factors mean for the kinds of items students can report on in the paper.
- **Steps and results:** Follow the four steps described in the section "A Four-Step Process for Using the Exercise" and write up the results the students found through doing the exercise.
- **Appendix**: Attach selected documents/samples from the corpus to include in appendices so students can direct readers to specific examples of findings (e.g., "As seen in paragraph four in Appendix A.") noted in the text of their papers.

Students can then use this resulting document as a mechanism for designing more effective documents for individuals from a particular culture. Again, review exercises such as this one are foundational, and instructors can modify or expand the nature of the exercise by selecting different cultures for examination and different kinds of materials (e.g., printed documents versus online sites) for review and analysis.

2.9 Conclusion

There are cultural differences in high- and low-context communication, in language, and in thought patterns that can influence the presentation of technical information in terms of structure, content organization, amount of information provided, and the relationship between text and graphics. The authors hope that this chapter can provide readers with an analytical framework for engaging in effective technical and professional communication across different cultures.

The authors also believe that the study of the automotive workshop manuals examined in the chapter can provide the following general guidelines for those working in intercultural technical communication contexts:

- Always consider the differences between high- and low-context cultures, in perception and thought patterns, in language, and in processing graphics when communicating between cultures;
- Do not assume that any organized approach adopted by one culture in presenting technical information can be adopted easily by another. Technical documents are usually less explicitly written and less finely detailed and structured in a high-context culture than in a low-context culture. The text and graphics can be presented in an elaborative way with more graphics and less text or in a redundant way with fewer graphics. A system or a component can be structured as an entity with interrelation and context in a culture characterized by synthetic thought patterns, but should be structured with specific and separate elements in a culture that prefers an analytical thought pattern;
- Cultural context plays an important role in information comprehension. As shown in the user study, the language character system, amount of common knowledge shared within a culture, and thought patterns may influence comprehension. For

example, information presented with more graphics might be well understood by people who grew up with pictographs; the information might be searched from a general overview to concrete parts in a culture characterized by a synthetic thought pattern

The findings regarding cultural differences in the structuring, content organization, and text–graphic relationships in technical documents have several implications for communicators working in intercultural technical and professional communication contexts.

When communicating with readers from a low-context culture, with analytical thought patterns, and a phonetic language, technical and professional communicators should structure a service manual and repair instructions with more detail. Technical information should be explicitly written to give the reader a more detailed description and background information about the objects in use. In addition, the contents of a system or a component should be organized with specific and separate elements in the functional description. Furthermore, a redundant relationship between the text and graphics should be considered in presenting the technical information. That is, the information presented in the graphics should also be concretely described in the accompanying text.

For readers in a culture characterized by high-context, synthetic thought patterns, and a pictographic language, it is suggested that technical and professional communicators structure the service manual and repair instructions with less detail. The technical information can be implicit since readers from this particular cultural background can generate the necessary information themselves based on the physical environment or personal connections. With regard to the content organization, a system or a component in the functional description should be structured as an entity with interrelation among its elements and context. It should also be noted that technical information should be presented with more graphics and the accompanying text should complement the graphics.

References

1. G. Hofstede, *Cultures and Organizations: Software of the Mind*. London: McGraw-Hill, 1991.
2. W. Fukuoka and J. H. Spyridakis, "Japanese readers' comprehension of and preferences for inductively versus deductively organized text," *IEEE Transactions on Professional Communication,* vol. 43, no. 4, pp. 355–367, 2000.
3. J. H. Spyridakis, "The effect of inductively versus deductively organized text on American and Japanese readers," *IEEE Transactions on Professional Communication,* vol. 45, no. 2, pp. 99–114, 2002.
4. D. D. Ding, "The emergence of technical communication in China - Yi Jing (I ching)," *Journal of Business and Technical Communication*, vol. 17, no. 3, pp. 319–345, 2003.
5. Q. Y. Wang, "A cross-cultural comparison of the use of graphics in scientific and technical communication," *Technical Communication,* vol. 47, no. 4, pp. 553–560, 2000.
6. F. Maletzke, *Interkulturelle Kommunikation*. Wiesbaden, Germany: Westdeutscher Verlag, 1996.

7. D. Slobin, "From thought and language to thinking and speaking," in *Rethinking Linguistic Relativity*, J. J. Gumperz and S. C. Levinson, Eds. Cambridge, MA: Cambridge University Press, 1996, pp. 70–96.
8. R. B. Kaplan, "Cultural thought patterns in inter-cultural education," in *Readers on English as a Second Language: For Teacher Trainees*, K. Croft, Ed. Boston, MA: Little, Brown and Company, 1980, pp. 11–25.
9. K. Maitra and D. Goswami, "Responses of American readers to visual aspects of a mid-sized Japanese company's annual report: A case study," *IEEE Transactions on Professional Communication,* vol. 38, no. 4, pp. 197–203, 1995.
10. O. C. Tzeng and N. Trung, "Cross-cultural comparisons on psychosomatics of icons and graphics," *International Journal of Psychology*, vol. 25, no. 1, pp. 77–97, 1990.
11. C. M. Barnum and H. L. Li, "Chinese and American technical communication: A cross-cultural companion of differences, " *Technical Communication*, vol. 53, no. 2, pp. 143–166, 2006.
12. M. Mirshafiei, "Culture as an element in teaching technical writing,"in *Proceedings of the Society for Technical Communication Annual Conference*, Arlington, VA: Society for Technical Communication, 1992, pp. 557–560.
13. E. A. Thrush, "Bridging the gaps: Technical communication in an international and multicultural society," *Technical Communication Quarterly*, vol. 2, no. 3, pp. 271–283, 1993.
14. E. T. Hall and M. R. Hall, *Understanding Cultural Differences: German, French and American*. Yarmouth, ME: Intercultural Press, 1989.
15. J. M. Ulijn and K. St.Amant, "Mutual intercultural perception: How does it affect technical communication? Some data from China, The Netherlands, Germany, France, and Italy," *Technical Communication*, vol. 47, no. 2, pp. 220–237, 2000.
16. Y. Q. Wang and D. Wang, "Cultural Context in International Technical Communication: A Study of Chinese and German Automobile Literature," *Technical Communication*, vol. 56, no. 1, pp. 39–50, 2009.
17. C. Wallin-Felkner, *Technische Dokumentation (in German)*. Augsburg, Germany: WEKA Fachverlag fuer technicsche Fuehrungkraft, 1998.
18. P. Gerigk, *Kraftfahrzeugtechnik*. Braunschweig, Germany: Westemann Schulbuchverlag, 1997.
19. F. Ganier, "Factors affecting the processing of procedural instructions: Implications for document design," *IEEE Transactions on Professional Communication,* vol. 47, no. 1, pp. 15–26, 2004.
20. Y. Jia, "Cognitive differences and intercultural communication: A field approach," *Ph.D. dissertation, Department of English Language and Literature*, Xiamen University, 2004, pp. 45–101
21. R. E. Nisbett and K. Peng, "A culture and systems of thought: Holistic vs. analytic cognition," *Psychological Review*, vol. 108, no. 2, pp. 291–310, 2001.
22. M. Wu, *Education Committee for the Worker of Automobile Transport, Structure and Repair of Automobile*. Shanghai: Shanghai Science and Technology Press, 1993.
23. S. J. Guan, *Intercultural Communication (in Chinese)*. Beijing, China: Peking University Press, 1995.
24. K. Seki, M. Chisaka, M. Eriguchi, H. Yanagie, T. Hisa, I. Osada, T. Sairenji, K. Otsuka and F. Halberg, "An attempt to integrate Western and Chinese medicine: Rationale for applying

Chinese medicine as chronotherapy against cancer," *Biomedecine & Pharmacotherapy*, vol. 59, Supplement 1, pp. S132–S140, 2005.

25. X. Zhou, Z. Wu, A. Yin, L. Wu, W. Fan and R. Zhang, "Ontology development for unified traditional Chinese medical language system," *Artificial Intelligence in Medicine*, vol. 32, no.1, pp. 15– 27, 2004.
26. S. P. Ballstaedt, "Bildverstehen, Bildverstaendlichkeit – Ein Forschungsüberblick unter Anwendungsperspective," in *Wissenschaftliche Grundlagen der technische Kommunication*, H. P. Krings, Ed. Tübingen, Germany: Gunter Narr Verlag, 1996, pp. 191–233.
27. C. Kostelnick, "Cultural adaptation and information design: Two contrasting views," *IEEE Transactions on Professional Communication,* vol. 38, no. 4, pp. 182–196, 1995.
28. Y. Q. Wang and Y. Z. Jiang, "Culture and Text-Graphic Relationship – A Study of Automotive Service Literature from China and Germany," in *Proceedings of IEEE International Professional Communication Conference*, Seattle, WA: IEEE Professional Communication Society. 2007, pp. 1–8.
29. Z. S. Wang, *Principle, Diagnosis and Repair of Electronic System in Vehicle*. (in Chinese) Beijing, China: Peking University Press, 1995.
30. D. Liang, *CA141 LKW Maintenance and Repair Handbook* (in Chinese). Changchun, China: Jilin Science and Technology Press, 1996.
31. F. M. Zahedi, W. V. van Pelt, and J. Song. "A conceptual framework for international Web design," *IEEE Transactions on Professional Communication,* vol. 44, no. 2, pp. 83–102, 2001.
32. Daimler-Benz AG, *Mercedes-Handbuch für die komplette Fahrzeugtechnik*, 1988.
33. E. A. Jones and C. Aoki, "The processing of Japanese Kana and Kanji characters," in *The Alphabet and the Brain: The Lateralization of Writing*, D. de Kerkchove and C. Lumsden, Eds. Berlin, Germany: Springer, 1988, pp. 301–320.
34. K. Guo, "Double coding and bi-hemispheres-character," in *Proceedings of the Seminar on Chinese Language Issues* (in Chinese). Beijing, China: Language and Literal Press, 1988.
35. S. H. Liu, "The use of analogy and symbolism in traditional Chinese philosophy," *Journal of Chinese Philosophy*, vol. 1, no. 3-4, pp. 313–338, 1974.

3

Teaching Image Standards in a Post-Globalization Age

Audrey G. Bennett

Rensselaer Polytechnic Institute

The emergence of open-access alternatives to industry-standard, image-making software enables non-designers (i.e., students in disciplines outside of communication design) to design professional images. The problem that exists is that non-designers (e.g., many individuals working in technical and professional communication) might not know what constitutes a communicatively effective image. *My previous work defines what constitutes such images from a communication design perspective. This work also provides metrics for assessing the communicative effectiveness of images. This chapter extends this work by situating those metrics within an interdisciplinary framework that includes technical and professional communication. The aim of this chapter, in turn, is to provide educators in technical and professional communication with metrics for teaching their students how to design communicatively effective images in a computer-mediated and borderless world.*[1]

---

[1] This chapter evolved from a previously published paper titled "Teaching design standards in a socially conscious age" in *Transformative Dialogues: Teaching and learning eJournal*, vol. 4, no. 2.

Teaching and Training for Global Engineering: Perspectives on Culture and Professional Communication Practices, First Edition. Edited by Kirk St.Amant and Madelyn Flammia.
© 2016 The Institute of Electrical and Electronics Engineers, Inc. Published 2016 by John Wiley & Sons, Inc.

3.1 Image Design and Consumption in a Post-Globalization Age

Over the past decade, technological innovations introduced two important phenomena that influenced current image-making practice [1]. First, technologies leveled the world. Second, they enabled the proliferation and widespread cross-cultural dissemination of propaganda regarding social and environmental issues like climate change and global warming [2].

3.1.1 The Changing Landscape

The technological leveling of the world had a profound effect on the design of images. Today, there is a high demand for images that both

- Address social issues
- Resonate culturally with diverse users

With the rise of awareness regarding social issues like global warming, for instance, it has become clear that humanity needs more images that serve a broader purpose in society beyond helping corporations to profit. There is a dire need for "communicative effective images that relay information to targeted users [to] foster a change in belief or behavior" [3] that leads to good social change. However, technical and professional communication students who seek to learn how to create professional images with social value will likely find themselves lost within an intellectual maelstrom of communication design perspectives that promote polar views on what constitutes a communicatively effective image. This chapter, however, purports that both perspectives are important to the advancement of humanity.

On the one hand, there are communication designers—like those associated with competitions like the Red Dot Design Award—who still primarily promote visually pleasing, corporate design as the highest standard. The Red Dot Design Award specifically targets entrants who want to "distinguish their business activities through design" [4]. On the other hand, there are communication designers (e.g., those associated with competitions like Sappi's Ideas that Matter) that advocate for design projects that address social and environmental issues (above aesthetic appeal) as representing the highest design standard. Sappi's Ideas that Matter competition provides funds to communication designers to use their expertise to further promote the mission of nonprofit organizations through communicatively effective images.

3.1.2 Agents of Change

Technological innovations in recent years also democratized the design of images. They enabled anyone with access to the World Wide Web to create professional images that communicate to a mass, global audience. This is a DIY/do it yourself or "design it yourself" [5], and DIWO/do it with others age of open-access to professional image design resources. With the click of a mouse, non-designers can upload images to professional Wordpress-enabled websites accessible to virtually anyone on the planet. Today, non-designers can also freely access and use Gimp, Inkscape, and Scribus—instead of their at-cost and proprietary alternatives within the Adobe Creative Suite—to design and propagate images globally by way of social media and networks. Moreover, these open-access tools are equivalent in functionality to the industry-standard software they imitate. As a result, today's playing field for professional image design has expanded vastly to include non-designers around the world.

With open-access, everyone can design. Yet this privilege that technological innovation affords has a drawback. First, and perhaps foremost, it exacerbates the following preexisting problem in communication design. That is, everyone now believes he or she is an expert, image designer. The question that arises then is this:

Can everyone design communicatively effective images?

While non-designers do indeed have ease of access to professional image making tools, the availability of standards for what constitutes communicatively effective imagery tends to be scarce. Furthermore, they are also often scattered, conflicting, and, at times, even inaccessible to those without high-tech resources such as high speed Internet access or smart devices.

The question I aim to address in this chapter is this:

How can educators in technical and professional communication teach non-designers to design communicatively effective images?

To answer this question, I build on my previous work on image standards and metrics [1]. I do so by using evaluation forms that technical and professional communication educators can use in their classrooms in order to objectively evaluate images and thereby relay to their non-design students how to communicate effectively with images. My goal is to guide technical and professional communication instructors in teaching students how to design social images (i.e., images that address social and environmental issues) that communicate effectively across cultural borders.

3.2 Socially Conscious Communication Design and the Evolution of Image Standards

During a recent experience teaching image design, I had an epiphany: The communication design discipline has a lacuna when it comes to collective standards for evaluating the communicative effectiveness of images. In applied communication design programs, design students are taught to be master aestheticians. According to this approach, students learn that the ultimate goal of the design process is to yield an outcome that pleases the senses—particularly the sense of sight. In fact, Paul Rand [6]—the late Paul Rand [6]—a leading communication design practitioner and educator—reveals the discipline's strong sentiment—reveals the discipline's sentiment toward visual aesthetics when he writes,

> To make the classroom a perpetual forum for political and social issues...is wrong; and to see aesthetics as sociology is grossly misleading. A student whose mind is cluttered with matters that have nothing directly to do with design...who is being overwhelmed with social problems and political issues is a bewildered student; this is not what he or she bargained for, nor, indeed, paid for. [6, p. 123]

Rand's strong bias in favor of aesthetics over social issues is a perspective that still resonates in many practice-based, communication design programs in the United States. In fact, this perspective yielded a core set of learning outcomes centered on attaining visually pleasing images that promote companies and sell their products. However, as recent history shows, what designers also want and need to sell are images that address social issues and, at least, aim to change the world for the better [7].

3.2.1 Approaches to Images as Tools for Social Action

Throughout history, we have seen images used as a tool for social activism around the world [8]. To serve nonprofit organizations, communication designers have created images for community bulletin boards, transportation vehicles, billboards, ads, and other public spaces. These images translate their client's social advocacy messages into visual rhetoric aimed at persuading others to adopt a new attitude, an alternative way of thinking or acting, or a different conscience toward unethical or environmentally damaging behavior.

A quintessential example of the historical contributions communication designers have made to the promotion of humanitarian and environmental issues in society is that of the work of the late communication designer Tibor Kalman. Kalman created, edited, and designed the magazine called *Colors* for Benetton. In each issue of *Colors*, Kalman employed visuals to advocate for various social issues including racism and HIV/AIDS. He used images that were arresting, striking, and rhetorically persuasive. As a result, Kalman's work in *Colors* epitomizes the harmonious amalgamation of visually pleasing aesthetics and social issues content that refutes Rands' implied claim that the two, like oil and water, do not mix.

3.2.2 Connections to Pedagogy

Communication designer Jorge Frascara is a well-known proponent to the inclusion of social issues content in communication design curricula. He is also a leading advocate for user research methods that contrasts with the intuitive approaches common in aesthetics-driven curricula in art schools. Frascara [9] argues

> I would hope that today, instead of trying to keep on inventing style superheroes in design shows, it would be refreshing to pay attention to social and cultural relevance, to the effectiveness of design solutions, and to the contributions that design makes to its highest possible function: supporting and fostering the welfare of people [9, p.18].

Frascara's work advocates collectively for exploring the role of communication design in addressing social issues. It also closely parallels the ideas presented in the *First Things First Manifesto* [10, 11] and other core literature on the subject of socially conscious design including works like Berman [12] and Heller and Vienne [7]. However, this chapter argues that the inclusion of aesthetics driven pedagogy married to social issues pedagogy that includes user research provides the ultimate potential for images to move beyond merely addressing social issues to effecting social change.

Communication designer Dietmar Winkler [13], another proponent of the inclusion of research in design pedagogy, argues:

> Looking at American design education, most schools curricula are rooted in the Bauhaus and its model of design with emphasis on formal aspects; hand and technical skills. Visualizing skills are still considered as most essential. This concentrated focus on visualization skills overshadows the student's intellectual and cognitive development. If design practice is to emerge as a recognized profession, it must overcome the barriers of the intellectually limiting Bauhaus model. Like other professions, it must begin to develop an information and knowledge base founded on contributions to a body of original research…The design practice must begin to recognize the need and to support the idea of research and testing. Design education must expand its…curricula to include…the introduction to research methodology and human, social and environmental factors [13, p. 9].

The underlying premise of Frascara's and Winkler's perspectives is that user research in the design process is a means to an end (the end being positive social change). This is because it requires testing the effect of images on target users and collaborating with them throughout the image design process. However, this inclusion of user research in the image design process as a way to effect positive social change requires a new communication design curriculum. Such a curriculum must marry the strong foundation of aesthetics-based teaching and learning championed by Rand and others with the social issues curricula advocated by Frascara, Winkler, and their successors.

3.2.3 Expanding to the Global Context

On a global/international scale, developing new design curricula that combine aesthetics-based teaching and learning with social change requirements is no small task. Rather, it warrants a new set of standards for what constitutes communicative effectiveness. Within this context, the Bauhaus legacy in communication design stressed visual appeal as the highest standard. Yet it also stressed the social consciousness movement that puts images that make the world a better place on the pedestal of communicative effectiveness.

In this chapter, I broaden the definition of a communicatively effective image as one that marries these two polar perspectives of images as agents for corporations versus society. I posit that images, today, should have aesthetic appeal and social change purpose. This new definition of what constitutes a communicatively effective image, however, needs standards.

3.3 Standards for Communicatively Effective Images

My previous work [1] delineates the following standards for images today:

- A communicatively effective image establishes credibility.
- A communicatively effective image stimulates use and facilitates ease of use in a public context.
- A communicatively effective image includes the user.
- A communicatively effective image resonates with the culture of users.
- A communicatively effective image sustains humanity or the environment.

3.3.1 A Communicatively Effective Image Establishes Credibility

When an image establishes credibility with the user, it is professionally presented, communicates ethical content [14], and establishes trust with the user through its form and function. An image that reflects these three aspects embodies various attributes that can be assessed. For instance, one can argue that most issues of *Colors* magazine have a unified aesthetic and sequence across its multiple pages. Each issue reflects good craft and professionalism in its presentation and uses the appropriate resolution for graphics for the target context of use. Its images reflect a high level of creativity and innovation. The typographic treatments are readable and legible [15, p.17][16, p.184][17, p. 63]. An underlying grid is present and used effectively to convey the information in orderly ways [18, p. 9] that are creative and accessible. There is a presence of hierarchy and contrast to organize information and guide the reader through the pages [19, p. 94][20, p. 172] [21, p. 316][22, p. 336]. The typographic treatments reflect an appropriate selection of fonts, point sizes, kerning, tracking, and leading. The information is sufficient, visually and verbally engaging, and accurate. The text- and image-based information is synthesized in the composition.

Regarding the verbal credibility of text within an image, two scenarios exist: Either the designer writes the information, or a client does. When the designer has control over the text development and its editorial process, then the text should be credible as well as the overall image that contains it. Writing reflects credibility through a quality argument that is sufficient, interesting, accurate, and positioned within an existing point of view. The author uses proper punctuation for text and citations; and, there are no spelling or grammatical errors. There are references to previous scholarship on the topic; and, supporting graphics are present and used effectively. For instance, images by the Guerilla Girls, an anonymous group of female activists, epitomize verbal and visual credibility.

3.3.2 Communicatively Effective Images and Use in Public Contexts

Existing usability perspectives focus primarily on testing the functionality of images like websites [23] and three-dimensional products [24] with their target user group. With such ideas in mind, usability testing can also benefit "static, two-dimensional images" [3] that aim to inspire the user to act—like change a life-threatening behavior. This chapter analogizes a communicatively effective image to a usable image that has the same characteristics as usable products and interfaces. A communicatively effective image stimulates use and facilitates ease of use.

I, however, posit that it also is a viable, well-timed solution that can be implemented within a public context and coalesce with existing social and political frameworks. Within its public context, an image facilitates use by possessing a clearly communicated, welcoming, and accessible point of entry and a comprehensible system of navigating the information it presents. The interface of the image, a billboard for example, attracts attention and guides users through a hierarchy of functions.

Ease of use is an uninhibited ability to access and interact with a tangible or intangible image and extract meaning. Its functionality is intuitive and requires little training beyond, perhaps, reading a brief manual (whereas how to use a billboard tends to be self-explanatory and intuitive). Interaction with a usable image effects pleasurable and memorable experiences. When people experience ease of use interacting with an image, then it is characterized as user-friendly. It communicates effectively.

To better understand these ideas, consider the following example: The images in rows two through four in Figure 3.1 represent a usable image designed through an iterative design process with extensive user testing.

If people experience the opposite, that is, difficulty accessing or interacting with an image, as in the top image in Figure 3.1, then they become frustrated or are unable to extract meaning; and, communication fails.

3.3.3 The Design Process of a Communicatively Effective Image Includes the User

A communicatively effective image comes from an iterative, collaborative design process—that is, a design process influenced by or inclusive of input from users and other stakeholders. Generally, when the goal is to attain visual appeal, an intuitive approach might suffice. However, when the end goal is to effect social change on a cross-cultural scale (caused by the flattening of the world as previously described), an approach that is inclusive of user input becomes necessary due to anticipated differences in culture between the image designer and target user group. A user research approach in a classroom setting might, in turn, occur simply with surveys of target users or through more rigorous (and costly), qualitative or quantitative methods (e.g., ethnography).

Image designers who conceptualize and develop their ideas in collaboration with their target users have a greater chance of their images resonating culturally with the target user group. To include users, one should employ a user research methodology that engages them in the image design process. For instance, the bottom left image in Figure 3.2 aims to communicate "Use a condom" to Ghanaian. Interestingly, a user testing session with Ghanaians disclosed that the image and its message were incomprehensible. All 25 Ghanaians tested could not decipher that a woman was whispering into the ear of her prospective male sex partner. Without user testing, I might have incorrectly assumed the image would clearly communicate that particular idea.

In some cases including the user in the image design process "makes [him or her] the localization expert" [25]. That is, the user becomes the person who controls how the image gets "adapted to a particular language, culture, and desired 'look and feel'" [26]. Yet visuals that provide user customization options that are unlimited might lead users

FIGURE 3.1. Act Now HIV/AIDS awareness and prevention campaign (See http://www.interactiveimage.info/). *Top*: The original printed image designed in a participatory manner with and for Kenyans in Kusa, Kenya; *Rows two through four*: The redesign of the original image with global users in the United States and elsewhere (through information and communication technologies (ICTs) for remote collaboration). The image on the *left in the second row* is the first of four screens that guide the viewer through a process of learning about the campaign and the functionality of an interactive version of the original image where viewers can customize the image to their cultural preference (*see the image on the right in the third row*) before disseminating it to their social networks or printing it for display via a local community bulletin board (via the screen on the *bottom*).

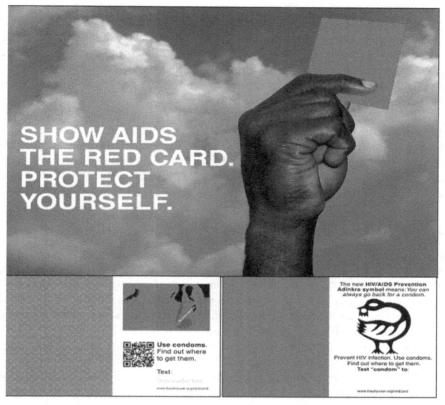

FIGURE 3.2. Show AIDS the Red Card HIV/AIDS awareness and prevention campaign used in Kumasi, Ghana. *Top*: Image designed in a participatory manner with Ghanains during the Summer of 2010; *Bottom left:* The red card referenced in *top* image, designed in the USA in 2011 and tested with Ghanaians in Ghana during the Summer of 2011; *Bottom right*: the next iteration of the red card designed in the USA in 2012 that includes an HIV/AIDS Adinkra symbol designed in a participatory manner with Ghanaians in Ghana Summer of 2011.

to design communicatively ineffective images simply because the users have too much control over final design decisions. Thus, there is still a dire need for the presence of an expert image designer in the user-centered, design process. The role of the expert would then be to ensure the image outcome meets the discipline's standards for visual and aesthetic appeal (among the other standards discussed in this chapter that are involved in communicative effectiveness).

3.3.4 A Communicatively Effective Image Resonates With the Culture(s) of Users

In addition to creating pleasurable and memorable experiences, images also have the potential to communicate meaning. Whether the image reaches its potential depends on its inherent ability to speak effectively to the cultural background and idiosyncrasies (e.g., preferences for colors, symbols, language, and navigation styles) of the target users. A communicatively effective image "avoids cultural stereotypes" [27,28]. It communicates through culturally appropriate graphics [29] that the expert image designer chooses in collaboration with the user, to resonate with the user's age; gender; ethnicity; context; mental, physical, and intellectual capabilities [23, p. 302][30, p. 288]; sociocultural and political values; and media literacy.

When an image resonates with the culture(s) of its users, it uses culturally appropriate aesthetics. These include being gender-fair; age appropriate; multicultural; mentally, physically, intellectually, politically and technologically accessible. For instance, the redesign of the red card on the bottom right in Figure 3.2 to display a new HIV/AIDS Adinkra symbol resonated better with the culture of Ghanaians and made the image more communicatively effective with members of that target audience.

3.3.5 A Communicatively Effective Image Sustains Humanity or the Environment

What is the lifecycle of the image? Does it reflect a "crade-to-grave" [31, 32] or cradle-to-grave approach? Does it contribute to the emission of greenhouse gases, the carbon footprint; or, does it sustain humanity and its environment? Has it been tested or evaluated in a public context? The answers to questions such as these are important to achieving the objectives of different stakeholder groups.

In essence, a communicatively effective image is disposable, recyclable, or adaptable to other uses. It reflects environmentally friendly choices like soy ink and recyclable paper. To be communicatively effective, the image has to embody both formal qualities and functionality that contribute to the sustenance of the environment and humanity. For instance, the interactive image in Figure 3.1 strives for communicative effectiveness through an eco-friendly, web-mediated form that aims to sustain humanity.

3.4 Implementing Objective Metrics in Technical and Professional Communication Classes

The use of metrics in the evaluation of images in the technical and professional communication classroom brings objectivity to a process that tends to be very subjective in communication design. In the communication design classroom, instructors typically rely on their own tacit standards for what constitutes communicatively effective imagery to evaluate professional exemplars and student work. Evaluation occurs within the context of a group critique that includes the instructor, students in the class, and sometimes invited critics who are typically professional image designers from industry.

3.4.1 Conventional Approaches to Evaluating Image Design

During most critiques of images designed by students, each student generally undergoes a group critique where he or she receives constructive feedback on the strengths and weaknesses of his or her image as well as suggestions on how it could be improved.

Comments from the student's peers tend to focus on the visual aesthetics (i.e., what did or did not look good) with shallow justifications that typically start with "I like it." or "I don't like it." After the critique, the instructor evaluates each student's image with a letter grade. This grade is often justified through subjective comments the instructor provides on the image's strengths and weaknesses and is generally done with disciplinary jargon that is sometimes incomprehensible to non-designers.

Over the years, I've observed that, during group critiques, non-designers who lack the necessary vocabulary in communication design find it difficult to participate orally in the critique. They don't know what to say. Their responses to grades under the subjective model typically range from apathy to outrage. With the latter emotion expressed frequently, there became an urgent need for a more objective approach to the evaluation of images. Thus, the question now becomes: What conventions or metrics can technical and professional communication instructors use to determine whether an image meets the standards for communicative effectiveness?

Table 3.1 addresses this question with a list of metrics gleaned from the standards in Section 3.3 of this chapter.

The metrics presented in this table aim to distinguish professional and expert images from amateur ones done by lay users and non-designers who have open access to the industry-standard tools for image design. Also, by including metrics that go beyond assessing aesthetic appeal to evaluating the degree of cross-cultural resonance, the global reach of images afforded by ICTs today gets considered in the image design process. The standards and corresponding list of metrics of Sections 3.3 and 3.4 are foundational to effective technical and professional communication education involving image design.

3.4.2 A Revised Approach to Image Evaluation

The metrics list presented in Table 3.1 facilitates a primarily objective evaluation of an image. It can be used to conduct formative evaluations by the student of his or her iterative image design process before design. The list empowers non-designers to self-assess their work before they submit a final version for instructor evaluation with the same metrics. Within the technical and professional communication classroom, I use the metrics list in conjunction with the following numerical scale to evaluate images according to their quantitative adherence to the metrics:

- 4: A score of 4 means excellent work. The image is missing 0 relevant conventions.
- 3: A score of 3 means good work. The image is missing 1–2 relevant conventions.
- 2: A score of 2 means average work. The image is missing 3–4 relevant conventions.
- 1: A score of 1 means the work needs improvement. The image is missing over 5 relevant conventions.
- 0: A score of 0 means poor work.
- NA: A score of "Not Applicable" means that a particular metric was not relevant to the image.

TABLE 3.1. Metrics for Evaluating the Communicative Effectiveness of Images

Communicatively Effective Conventions

Credibility

- The text uses supporting images appropriately and effectively.
- There is a synthesis of all graphics and text in the image's composition.
- The information is visually organized and coherent with a unified appearance and sequencing (when appropriate).
- The information is aligned and organized effectively according to an underlying grid.
- The aesthetic treatment of the image's layout stimulates and facilitates use cross-culturally.
- There are no decorative-only images or aesthetics.
- The image exhibits a high level of creativity or innovation.
- The image is designed and presented professionally in high-resolution (print: 300 dpi color, 150 dpi grayscale; screen: 72 dpi).
- Only 1 to 2 fonts are used.
- Column widths are appropriate.
- There is an appropriate selection of font(s), point size(s), kerning, tracking, and leading that effects readability and legibility.
- The image is accessible by way of the targeted senses (e.g., the image is legible and audible).
- The information is sufficient in quantity and accurate.
- One may consult with a copyeditor and proofreader as needed to satisfy the following metrics:
- The argument of the text is thoughtful and expressed within a theoretical framework.
- The text reflects great depth of knowledge (i.e., it includes high-quality observations, analyses, descriptions and reflection).
- No errors in punctuation, spelling or grammar exist.
- No errors in word usage, subject and verb agreement or sentence structure exist.
- When appropriate references to key literature (e.g., evidence-based research) are present and in the correct format.
- The text demonstrates logical sequencing of ideas through well-developed paragraphs; transitions are used to enhance organization.
- The text is well written, concise, and clear and stays on topic.

Ease of Use

- The image has a user-friendly interface that facilitates and stimulates use.
- The image's proposed context of use is viable and accessible by the targeted user.
- The image incorporates appropriate materials for the context of use.

User Input

- The image is designed through an iterative process inclusive of user input (with IRB approval as needed) and other stakeholders.
- The image's consumption process integrates user participation in an inter-sensory, interactive way that facilitates access by global, target end-users.

TABLE 3.1. Metrics for Evaluating the Communicative Effectiveness of Images

Communicatively Effective Conventions

Cross-cultural Resonance

- The image uses culturally appropriate aesthetics that respect and acknowledge the user's gender, age, impairment, literacy, etc…
- The image communicates in a way that resonates with the culture(s) of its users.
- The image displays a clear ethical sensibility that shows respect for the user, designer, and society (including the environment).

Sustainability

- The image shows potential to make a positive social impact (student images only).
- The image makes positive social change with a measure that shows statistical significance (professional images only).
- The image uses eco-friendly materials (e.g., materials that are energy efficient).
- The image is sustainable.
- The image can be adapted for other uses or recycled. It follows cradle-to-cradle lifecycle rather than cradle-to-grave.

Before submission to the instructor for final evaluation, students present their images to the class in a group critique. During the group critique, each student uses pins or tape to post their image onto a display board or digitally display it from a computer onto a screen by way of a projector. The metrics list provides a kind of shared vocabulary between the design and non-design students and all the students and me during the critique.

Though the use of metrics did not eliminate shallow feedback, the use of metrics did substantially decrease the use of the vacuous phrases "I like it." and "I don't like it." It is refreshing to hear students use the vocabulary in the metrics to critique the work of their peers. They'd use comments like "The text is not readable or legible" and explain how the typographic treatment could be changed to communicate better. The metrics effected broader discussions of an image's worth—beyond merely how the image looks to its potential positive impact on society and resonance with the culture of its target users. After final submission of each image assignment, technical and professional communication instructors can use the same metrics to conduct a summative evaluation of the student's image.

In my own evaluations with the list of metrics, I categorize them into two categories: visual and verbal conventions—though I acknowledge the possibility for an image to be visual only, without verbal material. Table 3.2 shows the metrics re-categorized into visual and verbal standards, and the evaluation forms in Figure 3.3 and Figure 3.4 include a third category called professionalism.

In my own classes, I use this category to evaluate the student's professional etiquette, that is, his or her ability to meet the deadline and follow the assignment guidelines. Both evaluation forms show how the metrics could be used in different formats. The first one,

TABLE 3.2. Metrics Categorized as Either Visual or Verbal Conventions

Visual Conventions:

- The text uses supporting images appropriately and effectively.
- There is a synthesis of all graphics and text in the image's composition.
- The information is visually organized and coherent with a unified appearance and sequencing (when appropriate).
- The information is aligned and organized effectively according to an underlying grid.
- The aesthetic treatment of the image's layout stimulates and facilitates use cross-culturally.
- There are no decorative-only images or aesthetics.
- The image exhibits a high level of creativity or innovation.
- The image is designed and presented professionally in high-resolution (print: 300 dpi color, 150 dpi grayscale; screen: 72 dpi).
- Only 1 to 2 fonts are used.
- Column widths are appropriate.
- There is an appropriate selection of font(s), point size(s), kerning, tracking, and leading that effects readability and legibility.
- The image is accessible by way of the targeted senses (e.g. the image is legible and audible).
- The image has a user-friendly interface that facilitates and stimulates use.
- The image's proposed context of use is viable and accessible by the targeted user.
- The image is designed through an iterative process inclusive of user input (with IRB approval as needed) and other stakeholders.
- The image makes positive social change with a measure that shows statistical significance (professional images only).
- The image uses eco-friendly materials (e.g. materials that are energy efficient).
- The image is sustainable.
- The image displays a clear ethical sensibility that shows respect for the user, designer, and society (including the environment).
- The image incorporates appropriate materials for the context of use.
- The image shows potential to make a positive social impact (student images only).
- The image's design process integrates user participation in an inter-sensory, interactive way during consumption.
- The image uses culturally appropriate aesthetics that respect and acknowledge the user's gender, age, impairment, literacy, etc...
- The image communicates in a way that resonates with the culture(s) of its users.
- The image can be adapted for other uses or recycled. It follows cradle-to-cradle lifecycle rather than cradle-to-grave.

Verbal Conventions:

- The information is sufficient in quantity and accurate.
- *One may consult with a copyeditor and proofreader as needed to satisfy the following metrics:*
- The argument of the text is thoughtful and expressed within a theoretical framework.
- The text reflects great depth of knowledge (i.e. it includes high quality observations, analyses, descriptions and reflection).
- No errors in punctuation, spelling or grammar exist.
- No errors in word usage, subject and verb agreement or sentence structure exist.
- When appropriate references to key literature (e.g. evidence-based research) are present and in the correct format.
- The text demonstrates logical sequencing of ideas through well-developed paragraphs; transitions are used to enhance organization.
- The text is well written, concise, and clear and stays on topic.

IMPLEMENTING OBJECTIVE METRICS

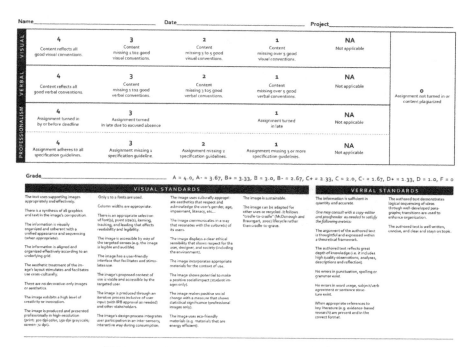

FIGURE 3.3. Sample student evaluation form 1.

Figure 3.3, places all metrics on the same level where each is of equal value to the other. However, the second evaluation form in Figure 3.4 evaluates each metric individually. Thus, there are opportunities to have one metric count more than another. The process of using each of these evaluation forms also differs. The evaluation form in Figure 3.4 is far more tedious and time-consuming during the grading process than the form in Figure 3.3.

For Figure 3.3, if a student received a 4 for a single category of metrics (e.g., visual conventions), that meant that he or she had mastered the set of conventions under that category. However, if a student received a 3, 2, or 1 that meant that there was one or more conventions not met by the image; and, the student need only look to see which were circled to find out which standards he or she did not master. Thus, an evaluation of 3, 2, or 1 meant that there was room for improvement; and, I encouraged the student to revise for a higher grade.

If a student received an overall unsatisfactory grade on any assignment, he or she had one week (from the time that he or she received her grade) to revise and resubmit. The use of this more objective procedure for the evaluation of student work in my classes effected greater resonance with students. It also generated more revisions since students knew exactly how to improve their work to get a higher grade.

Also, stating which evaluation metrics address a particular assignment in advance served to empower the students. That is, knowing what was expected at the beginning of

FIGURE 3.4. Sample student evaluation form 2.

the design process engendered a sense of control in students over their evaluation process. The use of metrics made students more accepting of their grades since they could clearly see which metrics they failed to meet. What is more important, the quality of student work improved with the implementation of objective evaluation metrics. Moreover, student evaluations improved as more students began to feel they had learned how to design images in the class.

3.4.3 Connecting Assessment Metrics to Teaching Practices

The use of assessment metrics by way of evaluation forms helps me focus each of my teaching activities toward addressing the standards for communicatively effective images. For each assignment in the class, for example, I tell the students which metrics the particular assignment or class activity addresses. I also note that the same metrics will be used to evaluate their work. The use of assessment metrics also clarifies the types of activities to include in the course curriculum. To teach:

- Visual credibility: I assign applied, image-design assignments where students either compose their own image or makeover a given image layout (e.g., a poster or a brochure) that visualizes another author's social issues content. Nonprofit organizations locally and non-governmental organizations abroad are good sources for this information.

- Verbal credibility: I assign applied assignments, like a codex, where students write about a topic and then use that content to design a book. This assignment is useful in intercultural and cross-cultural communication courses that might comprise much reading about culture in the United States as well as abroad.
- Ease of use: I assign a project where students use a set of usability heuristics to guide their analysis and redesign of an existing image in the real world. This assignment is particularly useful for existing images that aim to help humanity. Medicine and other types of informational labels or instructional manuals for social products are good examples of images that suit this assignment.
- User input, cross-cultural resonance, and sustainability: I assign a user-research project where students work collectively in small groups and in collaboration with members of a target user group to derive a solution to a social issue. Students must apply for IRB approval before working with their human subjects. Particularly for the sustainability standard, for images to effect social change that sustains humanity or its environment, users should participate equally in generating the problem and innovating a solution. An image designer can marry form and function with the intent to cause social change. However, to do so requires a less intuitive and more research-oriented, collaborative approach. And, for those who may not feel self-empowered with access to users abroad, ICTs can be used to engender agency and facilitate access to global users.

Through class discussions, the metrics can serve as an effective mechanism for examining issues of culture and international communication within the context of a technical and professional communication class. Discussions of assigned reading and media exemplars throughout the semester also allow me to demonstrate how to use the metrics presented in this chapter to analyze existing professional exemplars.

3.5 Conclusion

The question of what constitutes a communicatively effective image has never been more pertinent. With tech-mediated communication across geographic borders more accessible due to the rise of social media use, we can no longer evaluate images solely on their aesthetic appeal. More robust design processes that involve collecting data from users and involving them more in the design process is needed to create images that not only look good but resonate cross-culturally with global users.

When used, the standards and accompanying metrics presented in this chapter can better prepare students for creating images for international users and contexts of use. They make students aware that images can be designed for people from other cultures and that collaboration with users is important to attaining communicative effectiveness. An image reaches its full potential to effect positive social change when it adheres to the aforementioned criteria that affirm that non-designers can be designers of professional images that are visually pleasing and relevant to the needs of society. Images are more than just visual translations of a corporation's information. Images can also address environmental, health, and other social issues and change the world for the better.

In the next decade, the fruits of design labor have the potential to improve society; however, society's ability to realize that potential will take the collective effort of both designers and non-designers. Although open-access alternatives to industry standard software provide non-designers with ease of access to the technical knowledge that they need to make professional quality images, non-designers today who engage in image-making for communication purposes need to master more than only technical skills. They need to understand the discipline's collective standards for what makes an image communicatively effective. This chapter serves that purpose.

Another reliable source for the state of the discipline and its educational goals and standards is Icograda's Design Education Manifesto designed at the end of every decade since 2000. It is an updated document from the International Council of Design that provides recommendations for communication design education internationally into the future, one decade at a time. It also includes essays on the present and future of the discipline from communication design scholars from around the world. The 2011 Icograda Design Education Manifesto [33] confirms the metrics presented in this chapter that assess communicative effectiveness.

References

1. A. Bennett, "Teaching design standards in a socially conscious age," *Transformative Dialogues: Teaching and Learning eJournal*, vol. 4, no. 2, pp. 1–6, 2010.
2. D. Guggenheim, dir., *An Inconvenient Truth*. Paramount, 2006. Documentary film.
3. A. Bennett, *Engendering Interaction with Images*. Bristol, UK: Intellect Publishers, 2012.
4. *Red Dot: About Us*. Available at http://en.red-dot.org/3539.html.
5. E. Lupton, *DIY: Design it Yourself*. New York: Princeton Architectural Press, 2006.
6. P. Rand, "Confusion and chaos: The seduction of contemporary graphic design," in *Design Culture: An Anthology of Writing From the AIGA Journal of Graphic Design*, S. Heller and M. Finamore, Eds. New York: Allworth Press, 1997, pp. 119–124.
7. S. Heller and V. Vienne, *Citizen Designer: Perspectives on Design Responsibility*, New York: Allworth Press, 2003.
8. E. Resnick, C. Maviyane-Davies, and F. Baseman, *The Graphic Imperative: International Posters for Peace, Social Justice and the Environment, 1965–2005*, Boston, MA: Massachusetts College of Art, 2005.
9. J. Frascara, "A history of design, a history of concerns," in *Graphic Design History*, S. Heller and G. Balance, Eds. New York: Allworth Press, 2001, pp. 13–18.
10. S. Garland, First Things First Manifesto. *The Guardian*, April 1964. Available at: http://www.designishistory.com/1960/first-things-first/
11. R. Poynor, "First Things First Manifesto 2000," *AIGA Journal of Graphic Design*, vol. 17, no. 2, pp. 6–7, 1999.
12. D. B. Berman, *Do Good Design: How Designers Can Change the World*, Berkeley, CA: New Riders, 2009.

13. D. Winker, "Design practice and education: Moving beyond the Bauhaus model," in *User-centered Centered Graphic Design: Mass Communications and Social Change*, J. Frascara, Ed. London: Taylor & Francis, 1997, pp. 129–136.
14. P. Nini. In search of ethics in design. *Voice: The AIGA Journal of Design*, 2004. Available at: http://www.aiga.org/content.cfm/in-search-of-ethics-in-graphic-design.
15. R. Bringhurst, *The Elements of Typographic Style*, Vancouver, Canada: Hartley & Marks, Publishers, 1992.
16. K. Clair, *A Typographic Workbook: A Primer to History, Techniques, and Artistry*, New York: John Wiley and Sons, 2005.
17. J. Craig, *Designing with Type: The Essential Guide to Typography*, New York: Watson Guptill Publications, 2006.
18. T. Samara, *Making and Breaking the Grid: A Graphic Design Layout Workshop*, Gloucester, MA: Rockport Publishers, 2002.
19. E. Lupton, *Thinking with Type: A Critical Guide for Designers, Writers, Editors, and Students*, New York: Princeton Architectural Press, 2004.
20. R. Williams, *The Non-Designer's Design Book*, Berkeley, CA: Peachpit Press, 2008.
21. J. T. Hackos and D. M. Stevens, *Standards for Online Communication*, New York: John Wiley & Sons, 1997.
22. C. Kostelnick and D. D. Roberts, *Designing Visual Language: Strategies for Professional Communicators*, Boston: Allyn and Bacon, 1998.
23. J. Nielsen, *Designing Web Usability*, Berkely, CA: Peachpit Press, 1999.
24. J. Rubin, D. Chisnell, and J. Spool, *Handbook of Usability Testing: How to Plan, Design, and Conduct Effective Tests*, New York: John Wiley and Sons, 2008.
25. C. R. Lanier, "Making the user the localization expert: Employing user-customization strategies in globalizing online content," in *Culture, Communication and Cyberspace*, K. St.Amant and F. Sapienza, Eds. Amityville, NY: Baywood Publishing Company, 2011, pp. 39–61, 2011.
26. D. Cyr and H. Trevor-Smith, "Localization of web design: A Comparison of German, Japanese, and U.S. website characteristics," *Journal of the American Society for Information Science and Technology*, vol. 55, no. 13, pp. 1–10, 2004.
27. W. Horton, "Graphics: The not quite universal language," in *Usability and Internationalization of Information Technology*, N. Aykin, Ed. Mahwah, NJ: Erlbaum, pp. 157–187, 2005.
28. P. J. Hager, "Global graphics: Effectively managing visual rhetoric for international audiences," in *Managing Global Communication in Science and Technology*, P. J. Hager and H. J. Scheiber, Eds. New York: John Wiley & Sons, pp. 21–43, 2000.
29. M. Coe, *Human Factors for Technical Communicators*, New York: John Wiley & Sons, 1996.
30. J. Tidwell, *Designing Interfaces*, Beijing, China: O'Reilly Media, 2006.
31. W. McDonough and M. Braungart, *Cradle to Cradle: Remaking the Way We Make Things*, New York: North Point Press, 2002.
32. Defining Life Cycle Assessment (LCA), "US Environmental Protection Agency." 17 October 2010. Web.
33. A. G. Bennett and O. Vulpinari, (Eds.) *Icograda Design Education Manifesto 2011*, Montreal, Canada: Icograda, 2011.

Societal Contexts

4

Linux on the Education Desktop: Bringing the "Glocal" into the Technical Communication Classroom

Brian D. Ballentine

West Virginia University

For many European countries, use of and reliance on open source software is standard procedure for private business and government entities. However, the European politics and outlook on open source adoption are not mirrored in the United States. Beyond the more obvious answers of Microsoft® and Apple® being US-based companies, US hesitation and even resistance to open source is puzzling. This chapter will explore what we can learn from Europe's successful open source adoptions and will offer strategies individuals involved in professional communication practices can borrow from those successful models.

4.1 Introduction

In many European and Asian countries, Linux®-based, open source operating systems for desktop computing are increasingly becoming a viable option for private business organizations, educational systems, and government entities. While it is difficult to obtain precise data on market shares for different operating systems, the estimations on Linux-based operating system usage worldwide fluctuate between 1% and 1.6% of the total global market [1, 2]. The European and Asian mindsets that treat open source operating systems as a suitable replacement for Microsoft's Windows 7, XP, or Vista

---

Teaching and Training for Global Engineering: Perspectives on Culture and Professional Communication Practices, First Edition. Edited by Kirk St.Amant and Madelyn Flammia.
© 2016 The Institute of Electrical and Electronics Engineers, Inc. Published 2016 by John Wiley & Sons, Inc.

are often not mirrored in the United States. As this chapter will discuss, Linux success stories in US educational settings belong mostly to server-side solutions where, for example, an email server running Windows® is replaced with Linux. Desktop success stories in the United States generally remain confined to small projects often provoked by economic necessity, or they begin as pilot programs that do not mature to oust Windows as a dominant operating system. Beyond the more obvious explanation of Microsoft being a United States-based company, the reason for the absence of wider spread Linux distributions running on desktop computers in the United States remains a complex convergence of technical, social, political, and economic issues. That is, the answer goes beyond the old fear, uncertainty, and doubt (FUD) sales adage espoused by proprietary vendors: "You will never lose your job going with our company." Such factors have important implications for professional communicators who either use such technologies to design communication products (e.g., reports, websites) or who design documentation relating to such technologies (e.g., user manuals, help sites). Moreover, as the teaching of professional communication is intertwined with understanding the technologies used to achieve these ends, such trends also have important implications for professional communication education in different regions.

This chapter will begin by briefly acknowledging the worldwide success of Linux in the server and enterprise solutions market as well as its growth in the handheld device market. This discussion will also acknowledge that there are Linux desktop success stories in the United States and will also examine the *how* and the *why* of those successes (and failures). The primary focus of this chapter, however, will be to explore what we can learn from Europe's successful large-scale adoptions of Linux—particularly in educational contexts—by focusing on an autonomous region of Spain that, until recently, relied on a customized version of the Debian Linux distribution. Specifically, Spain's Extremadura region developed and supported "LinEx" for many of its government agencies and for its entire educational system. Such an examination, in turn, can provide professional communicators with a framework for considering how technology can affect professional communication practices in different international contexts.

The idea I wish to explore by examining these issues is that the *how* and the *why* of the LinEx success are tied to its meshing of the global and the local into what sociologist Roland Robertson deemed the "glocal" [3, p. 29]. By examining such ideas, I will reveal that LinEx's downfall in terms of gaining greater global market share is not due to technical or even socio-cultural deficiencies, but is largely the result of political circumstances—factors that can affect the use and adoption of most technologies (and thus, associated professional communication practices) in global contexts. In reviewing these topics, I will present specific educational strategies professional communication instructors can use to bring these Linux success stories. I will also discuss how to integrate these ideas into the professional communication classroom in order to help students better understand how the notion of combined global and local (or glocal) can affect international markets for and uses of technologies.

4.2 Linux—Dominance and Absence in Different Markets

In a recent and sobering post on his ZDNet hardware blog, Adrian Kingsley-Hughes tells his readers that if they are still waiting for the year of the Linux desktop, they should stop. That year simply is not coming. Kingsley-Hughes claims that Linux is "stuck permanently at a 1% market share. And that's where it'll be ten years from now" [4]. A Gartner report supported his forecast stating, "Linux OS is expected to remain niche over the next five years with its share below 2 percent" [1]. That report also predicted the continued dominance of Microsoft Windows and its various versions with Windows 7 leading with a total worldwide install base of 42%. At the end of 2015, Windows 7 held a share of more than 57% with Windows products dominating more that 91% of the total desktop and laptop markets [2]. The Linux prediction proved quite accurate at 1.6% at the end of 2015 [2]. While the strength of the Mac OS still "varies greatly by region," Mac was projected to grow its global market share of shipments on new PCs from 4.5% to 5.2% by 2015 [1]. But, by the end of 2015, the Mac OS was being used on just over 7% of desktops and laptops [2]. The numbers, however, appear much more favorable for Linux in the global markets for server and for handheld devices.

4.2.1 Linux and Server Markets

When it comes to business or enterprise-level servers, Linux powers many of the world's most popular companies from Amazon, to Google, to eBay, to Facebook, to Twitter [5]. Linux is behind many of the world's stock exchanges and has even found a role in the U.S. Department of Defense. Indeed, at a 2007 conference hosted by the Association for Enterprise Integration, Brigadier General Nick Justice is on record stating, "It may come as a surprise to many of you, but the U.S. Army is "the" single largest install base for Red Hat Linux. I'm their largest customer" [6]. Arguably the involvement of companies like Red Hat®, IBM®, Novell® and SUSE® (the last two both now owned by Attachmate) drives and then continues to support Linux's growth in the server market. That is, these companies have collectively spent millions of dollars on supporting open source development, so they may commercialize their customizable brand of Linux. These support providers, in turn, make a profit by charging customers for assisting them with migrating their systems away from Windows as well as continued support for those systems after migration is complete.

4.2.2 Linux and Handheld Devices

The Linux presence in the handheld and smart phone markets continues to grow due, in large part, to Google's Android operating system. The Linux kernel powers the Android OS that, according to recent data, now dominates the mobile market. Another Gartner report notes, "The Android OS accounted for 52.5% of smartphone sales to end users in the third quarter of 2011, more than doubling its market share from the third quarter of 2010" [7]. In a fashion similar to the way in which Red Hat, IBM, and others foster the growth of Linux in specific market sectors, Google and its Open Handset Alliance help foster the use of Linux in the handheld market. The Alliance is comprised of more than 80 companies ranging from mobile carriers, handset manufacturers, and software companies working with the open-sourced Android OS [8].

Google acquired Android in 2005, and by 2007, the company had not only released a software development kit (SDK) for Android, but it was sponsoring competitions for open source developers to design the best apps for the operating system. Cash prizes ranged from $25,000 to $275,000 for a total of $10 million dedicated to the contest [9]. Interestingly, Google does not have licensing fees for Android. Moreover, it does not take a percentage of the market sales associated with Android. The focus, instead, seems to be on achieving a degree of access through these devices—access that can be used as a foundation for connecting to consumers. For example, Google estimates that it "has activated some 190 million Android devices worldwide" and reportedly made $2.5 billion in ad revenue last year just from their mobile market [10]. Thus, access has its advantages. These numbers are particularly impressive when one considers they are connected to a software program that is licensed with the GNU General Public License (GPL) versus the more traditional, restrictive licensing practices associated with vendors of proprietary technologies (e.g., Microsoft). Thus, an understanding of such licensing practices can help us better understand perspectives associated with the adoption and use of more open source technologies such as Linux.

4.2.3 Linux and Licensing

Licensing practices are one of the most interesting aspects of open source software. Standard copyright licenses reserve all rights for the copyright holder and, by their very nature, do nothing to promote information sharing (in this case, the sharing of computer code) or knowledge distribution across users. That is, even if other developers are able to gain access to and read copyrighted code, the law prohibits non-copyright holders from reusing or repurposing that code for use in other applications. In the case of open source, a license is often referred to as a "copyleft" as opposed to a copyright license. The primary tenet of such copyleft licenses is to protect the freedom of individuals to use the software for any purpose, the freedom for individuals to change the software to suit a user's needs, the freedom to share the software with others, and the freedom to share the changes made to the software by other developers and by users of that technology.

The most common copyleft license employed by the open source community is the General Public License (the GPL is now in version 3.0), which specifies that changes made to software licensed under its tenets cannot be closed off or made proprietary. (Proprietary software is packaged in such a way that users cannot see the underlying computer code that allows the software to work; thus, individuals cannot "borrow" and modify such code to create other software products.) Finally, and of extreme importance to companies like Red Hat and IBM, the General Public License does not restrict the freedom to sell the software. If, for example, Linux used standard copyright licensing instead of the GPL, any company or individual wishing to legally use the software would be required to formally license (i.e., pay the copyright holder) Linux first. Additionally, the services and support offered by these commercial companies (e.g., Red Hat that sells Linux) must be exceptional as well as offer cost savings if their customers can otherwise download GPL-licensed software like Linux for free. Indeed, a user may download not just the compiled and executable version of the software, but the underlying source code as well.

This openness allows the user to "see" how the software works. It also allows the user to borrow some or all of this code as the individual sees fit and to use that code for other projects as needed or desired. In this way, open source software allows individuals to be both consumers and developers of a given product. GPL's ability to provide legal access to this source code, in turn, drives community contributions to the code and keeps open source initiatives like Linux alive and moving forward. In essence, users can become developers who can contribute to or modify software code. Through such activities, the user becomes part of a greater community dedicated to using and modifying the software.

4.2.4 Linux and Open Source Community Practices

The practices that have emerged from and around free, community contributions to open source code like Linux have caught the attention of researchers from a wide range of disciplines, including business and economics, sociology, political science, and professional and technical/professional communication [11–14]. Most of that scholarship addresses a broad range of changes in our social and cultural practices as they intersect with software development and use. For example, in *Two Bits: The Cultural Significance of Free Software*, Chris Kelty notes a variety of these changes including: "practices of sharing source code, conceptualizing openness, writing copyright (and copyleft licenses), coordinating collaboration, and proselytizing all of the above" [15, p. x]. In short, these changes have collectively led to the sustained growth of Linux, a sophisticated piece of free and community-developed software that is now at the heart of multimillion and even billion dollar global businesses.

The reasons, then, for Linux's impressive position within the server and handheld markets go beyond just the technical maturity and stability of the operating system. According to Kelty, "There are explanations aplenty for why things are the way they are: it's globalization, it's the network society, it's an ideology of transparency, it's the virtualization of work, it's the new flat earth, it's Empire. We are drowning in the *why*, both popular and scholarly, but starving for the *how*" [15, p. x]. Here I agree with Kelty, but only to a point. I believe early works such as Eric Raymond's now classic *The Cathedral and the Bazaar* offer at least a beginning discussion on the *how* of open source success. In Raymond's piece, he examines the open source email application Fetchmail. Focusing on a specific application enables Raymond to expound on his theories of how open source's bazaar-style approach to software development that emphasizes free-flowing information can outperform a proprietary or traditional cathedral-style approach to development [16].

In Kelty's defense, his project seems to be going after some of the larger ramifications regarding the cultural significance of free software practices. That is, how can those practices move beyond software and be employed to the betterment of our total human existence. For my purposes, in order to understand the *how* of the successes and failures of Linux as a desktop operating system, it is also necessary to contextualize this global, open source phenomenon within a specific local setting. By beginning with a specific Linux distribution (Spain's LinEx) we may then be able to pursue, as Kelty suggests, implications both local and global in order to best understand open source and what it may offer desktop computing in educational settings. Such an examination, in turn, can provide professional communicators with insights on how such technologies are adopted and used in different contexts and thus provide ideas and frameworks for selecting technologies and planning technology uses associated with different professional communication practices in international environments.

4.3 Linux on the Desktop

To help explore these ideas and issues, I will focus on Linux's desktop market share as well as explore the *why* and the *how* of Linux's modest successes in this market within educational settings. In essence, the year of the Linux desktop has arrived for very few people worldwide. However, as I will demonstrate, understanding the how and the why of those arrivals and departures offer enormous benefits to professional communicators as we attempt to navigate the technological facets of a growing global landscape.

4.3.1 The Global Market

Given all of Linux's other successes, why does it maintain less than 2% of the global market share of desktop operating systems? Why isn't the year of the Linux desktop coming for the other 98%? Kingsley-Hughes is dismissive of Linux's prospects remarking that the desktop market is "stitched up" with practices connected to original equipment manufacturers (OEM). That is, the designers of desktop systems often supply consumers with computers that has software (mostly Windows, but to a smaller extent, Apple's) pre-installed on it [4]. Simply put, some of the why and how of the desktop market are tied to business contracts and economics of the computer market.

These practices greatly affect the abilities of outsiders to introduce their products to prospective consumers. And the costs associated with moving from such pre-installed software to a competing product can create a barrier that prevents consumers from exploring and adopting other kinds of software. A Gartner report, for example, says we won't see strong Linux numbers because of the, "remaining high costs of application migration from Windows to Linux" [7]. Even as switching operating systems becomes easier, a company has to consider the operation and licensing of *all* the tools its employees rely on to be productive. Certainly more applications can run on Linux than in the past, and more computing is shifting to the platform-neutral "cloud." However, the uncertainty surrounding the viability of Linux to run all business application demands, in combination with the total cost of ownership, makes the migration appear too risky. That is, the often-cited concern over proprietary "vendor lock" (i.e., one company controlling or having a "lock on the market") is not enough to carry the day for a mass Linux migration on the desktop.

For Linux desktop proponents, what makes the Gartner report and similar forecasts so frustrating is that all of the previously named issues (vendor lock, application migration, OEM collaboration, and total cost of ownership savings) *are* actively being addressed for the business desktop. Perhaps the most notable effort is the partnership between billionaire entrepreneur Mark Shuttleworth's company Canonical® and the Ubuntu® Linux distribution. Ubuntu's "Desktop for Business" web page (see http://www.ubuntu.com/business/desktop) touches on all of the afore-mentioned points. It also claims that there is now a "real alternative" for the "endless cycle of Windows upgrades" with a product mature and stable enough "for large-scale corporate deployments of Ubuntu Desktop." Moreover, amidst the uninspiring prognostications for Linux on the desktop, is the site's exceptionally bold claim that "Ubuntu is used as a desktop operating system by thousands of businesses on millions of desktops around the world" [17]. While the casual use of "thousands" and "millions" does not exactly inspire a sense of statistical accuracy, we should acknowledge that even a 1–2% share of total worldwide computer users means that a significant number (not percentage) of people are using Linux as their desktop OS. To put it most ineloquently, those numbers are "not nothing."

Proof, too, is easier to come by with case studies starting to pile up online. Ubuntu's site, for example, offers several desktop success stories including Spain's Adalusia region and its deployment of 222,000 machines running Ubuntu as the desktop OS [18]. More modestly, the Canonical site reports saving the government of Kerala, India, tens of thousands of dollars on licensing fees by outfitting "141 members of its legislative assembly" with laptops running Ubuntu. Similarly, the Canonical/Ubuntu desktop solution saved Saklica City Hall in Slovakia more than 70% on its IT costs. According to news articles and press releases, over the past 8 years, schools in Germany [19], Russia [20], India [21], and South Korea [22] have all made major migrations to running versions of Linux on their school desktops. Thus, globally, open source desktops have made significant inroads within government and educational settings. These inroads increase the chance of students interacting with open source desktop computing and therefore increase the demand on instructors to adjust curriculum to include open source experience. It is developments such as these that professional communicators need to be aware of and understand, for they can play a central role in terms of the technologies one can use to share ideas and information with wider, global audiences.

4.3.2 Open Source in Educational Contexts

In the United States, Linux desktop success stories and case studies exist, but they are more difficult to come by. They typically occur on a smaller scale. Oakland, CA, elementary school teacher Robert Litt, for example, has won attention for building a computer lab from donated computers on which he runs only Ubuntu Linux and open source software. Litt claims he has had very little difficulty acquiring older (minimum of 512k of RAM) but still functional computers from local government agencies, businesses, and e-cyclers [23]. Quite often, according to Litt, organizations donating computers will claim that the machines are "broken." Litt says that a majority of the computers have software issues and not hardware problems. As a result, when he installs Ubuntu on such computers, they are "brought back to life."

Litt's first lab at the ASCEND K-8 elementary school started with 18 computers. Today, his Oakland school has a total of 70 donated machines running Linux. It is important to note, however, that Litt will be leaving ASCEND for a different teaching position [24]. One cannot help then but wonder if the lab will continue on—or will continue on in the same way—without Litt's presence. As this chapter will note, the sustained success of open source initiatives like Litt's computer lab often hinge on the dedicated efforts and leadership of a single individual. Scholars and open source advocates interested in studying Linux on the educational desktop will thus want to revisit the ASCEND program in the years to come in order to see how changes in staffing might affect this program.

Another example of open source use in education comes from a case study posted to SchoolForge (a site with an obvious nod toward the open source portal SourceForge) profiling the *Greater Houlton Christian Academy* in Houlton, Maine, that abandoned Microsoft entirely in 2002 [25]. That case study was updated with new information verifying the school's commitment to open source in 2011. The school's systems administrator, Michael Surran, also wrote a profile of his school in a 2008 issue of the *Linux Journal*. In order to make Linux work for all of his school's needs, Surran has had to go so far as to code his own educational applications, primarily in Python [19]. Many of the comments at the end of Surran's 2008 article point out that the success of projects like these is often due to the dedicated efforts of an invested Linux advocate, like Surran, who cares not just about cost, but also about user experience. According to the school's web site, his user base is small. Within their K-12 program, each grade level is capped at 20 students; so ultimately, this customized and localized Linux environment serves a few hundred teachers and students. The user base might be small, but without Linux as an option these students would receive no computer literacy training.

Given that Litt's and Surran's schools have basically no software budget and they get by on donated machines, their abilities to meet so many of the student and faculty computing needs no doubt count as genuine success stories. The sustainability of those open source projects remains to be seen. Other US educational Linux desktop projects have, despite initial successes, been abandoned for Windows. Linux desktop projects, especially the larger scale projects, are often pilot projects or projects funded temporarily by grants rather than wholesale adoptions and migrations. For example, Indiana public schools garnered a great deal of media attention in 2005–2006 when, as part of a government ACCESS (Affordable Classroom Computers for Every Secondary Student) grant, its administrators decided to deploy the Linspire Linux distribution on 24,000 school desktops. At the 2006 National Education Computing Conference, the project was lauded as "the largest K-12 desktop Linux deployment in the United States" [26]. Then Indiana Department of Education Special Assistant for Technology, Mike Huffman, gave a presentation at the conference touting cost savings and ease of use.

But, that was several years ago, and so much has changed. Indiana's chosen Linux distribution, Linspire, got its start as the now infamous "Lindows" distro (after a series of lawsuits Microsoft paid to acquire the Lindows trademark and Lindows changed its name to Linspire). In 2008, Linspire was acquired by Xandros, a Linux-based server and desktop solutions provider [27]. The Xandros site has not had updates posted in its news and events sections since 2009. The Xandros desktop forum, although not disabled, contains postings that debate the possible reasons for Xandros's discontinued pursuit of their Linux desktop OS under the forum thread, "After Xandros, then what?" Speculation ranges from business agreements formed with Microsoft, to the piece of proprietary code Xandros wrote for their file manager, to the simple conclusion that Linux on the desktop just isn't profitable [27]. It was also reported that Indiana's ACCESS program had contracted with a local company, Wintergreen Systems, to supply the computers with the Linux OS [28]. A visit to the Wintergreen web site reveals that the company now only officially vends Microsoft products, and there is no mention of Linux on the site. (Incidentally, in 2009, Wintergreen was formally charged by the Federal Trade Commission for defrauding thousands of customers of rebate checks they were supposed to receive after purchasing a variety of electronics from Wintergreen. The company was cited as a repeat offender [28].) Finally, Mike Huffman's LinkedIn page shows that he left Indiana public schools in January of 2009 to join an open source consulting agency called schools4tomorrow.org.

My attempts to contact Huffman and Wintergreen for comment on this case were unsuccessful. I was, however, able to interview the new IT manager for Southern Indiana's Education Center, Chad Smith [29]. He told me that he had not taken down the official Open Source Indiana web site supporting the defunct Linux ACCESS project because the page still contained useful links for individuals interested in pursuing open source projects and funding applications. Smith estimated that at the project's peak in 2008–2009, between 60% and 65% of their student desktops were running Linux. Today, Smith estimates that they are now running Linux on 10–15% of their desktops with the rest having reverted to Microsoft. Why and how did they scale back so quickly to Windows?

So what does all of this mean for educational practices relating to professional communication? In essence, integrating open source options into classes is not particularly difficult, nor is it necessarily costly. (Remember, most of the machines for Litt's aforementioned lab were donated.) They key is dedication and motivation. That is, instructors need to be dedicated to teaching using such alternative software/programs and might need to take it upon themselves to acquire the hardware and software needed to provide students with exposure to open source environments. Yet, as noted in the cases discussed here, getting the necessary technologies in place is only the first piece of a greater puzzle. Sustained commitment makes up the remainder, so individuals who wish to integrate open source options into their classes need to create not only plans to develop effective open source learning experiences/environments, but also plans to sustain the operations and uses of such environments over time. While such undertakings are no small task, they are not insurmountable, as Litt's and Surran's examples reveal, and in the end, they can both provide students with valuable learning experiences and reduce costs associated with providing professional communication students with the foundational technology skills needed to succeed in the modern workplace—skills that transcend the notion of particular software programs or specific platforms (e.g., understanding how to use design programs to create graphics or to use editing software to create videos).

4.3.3 Competition for the Education Market

To say that Microsoft aggressively protects its desktop products—particularly in an educational context—couldn't be more of an understatement. With Linux eating into the server and handheld device market, Microsoft works hard to maintain and grow its desktop market share. Incidentally, Smith's accounting of where Linux is used in Indiana schools coincides with national trends. He runs Linux for the school system's mail and web servers and uses applications like Moodle® and Joomla® for supporting dynamic learning environments and content management systems. As for the desktop, he said, Microsoft now offers significant educational pricing discounts and works with schools to customize offerings and accommodate their needs in ways they had not in the past. For Smith, the pricing and flexibility effectively makes Linux's total cost of ownership argument a moot point. Pair Microsoft's new customer approach with the loss of Indiana's in-house Linux advocate, Mike Huffman, and it isn't too difficult to imagine

Indiana's remaining 10–15% of open source use in the education area fading over the next few years.

Microsoft has been steadily growing its reputation and track record of working with universities so they may more affordably run and maintain Windows desktops. For example, in his book *Multiliteracies for a Digital Age*, Stuart Selber comments on his own university, Penn State, and its relationship with Microsoft:

> The upside of the Penn State-Microsoft Program is incontrovertible: at educational resale costs, the free software still amounts to a windfall of well over $500; teachers can make reasonably safe assumptions about software access on and off university grounds; and the extensive tutorials and workshops have been built up with centralized resources to educate and support students. [30, p. 121]

Indeed, the relationship appears to be a genuine boon for students and teachers. However, Selber makes it clear that a tuned critical literacy offers a more informed view. Specifically, he worries that passively accepting Microsoft products shuts down "critical comparisons between Microsoft and non-Microsoft software programs created for similar purposes" [30, p. 122]. In many respects, Indiana's ACCESS grant funding Linux on school desktops was designed to push such comparisons. Smith told me that in the early stages of the Linux deployment, his IT support staff's first taste of Linux was a "bad taste." Today, Linux offerings like Ubuntu can be easily managed for large-scale deployments but with relationships like the Penn State–Microsoft Program and the discounted educational rates Indiana schools receive, there is little incentive to go back to a Linux offering and compare.

In effect, an argument could be made that the single biggest barrier for Linux desktop adoption in US schools is that nobody is paying attention. Again, as Kingsley-Hughes points out, "Many consumers, in particular those who do most of their computing through the browser, would be more than happy with Linux... if they knew about it" [4]. And, even if users knew about it, where would they go for hardware running Linux? For example, if readers of this article were unaware of Canonical and Ubuntu's desktop initiative there is also a great chance that they have not heard of System76, ZaReason, Emperor Linux, Pogo Linux, or other OEMs that vend Linux. In a word, Linux desktop could use some evangelism.

4.4 Aggressive Evangelism

In her recent article, "The Technical Communicator as Evangelist: Toward Critical and Rhetorical Literacies of Software Documentation," Jennifer Maher writes, "Undoubtedly, the use of *evangelism* to frame an analysis of the discourse of the software development culture, which was founded upon the scientific principle of the algorithm, might strike many readers as forced at best and misleading at worst" [31, p. 370]. But, as Maher reveals, evangelism not only serves as a useful frame for her cogent analysis of open source software and its documentation; in fact, evangelism (not just mere marketing) is a built-in practice at proprietary companies like Microsoft. Maher points us in the direction of the *Comes versus Microsoft* case where a "highly confidential" internal document titled *Effective Evangelism* was submitted as evidence [31, p. 371]. In it, we get a sense of the company's aggressiveness and exactly how they define evangelism: "Evangelism is the art and science of getting developers to ship products that support Microsoft's platform" [32, p. 6]. The document also defines what the company considers as "victory." Specifically, "'A computer on every desk and in every home, running Microsoft software.' This is the mission statement of Microsoft itself; it is the definition of the conditions under which Microsoft itself can declare overall victory" [32, p. 5]. Linux victories, if they are to arrive for the educational desktop, will require an equally aggressive approach. For professional communication instructors interested in exploring and gaining support for open source options in teaching, an understanding of such evangelical processes and approaches can help with developing strategies related to advocating for open source options in relation to teaching.

4.4.1 Approaches to Evangelism

While Microsoft's quest for domination of the desktop has an all-encompassing or global objective to it, it is also quite clear that the company works on case-by-case, local levels with individual school districts (Indiana) and individual universities (Penn State and many others including my own university). Their methods, that is, their *how*, at the level of the local are impressive. According to a recent *New York Times* article, many companies, including Microsoft recruit campus representatives or what are now referred to as "brand ambassadors." Microsoft's reps give "interactive product demonstrations each week to peers on more than 300 campuses" [33]. The effectiveness of such campaigns is understood to be so powerful that specialized marketing agencies have cropped up that can manage a company's campus presence. For example, Mr. Youth, a marketing agency based in Manhattan, "charges corporate clients $10,000 to $48,000 a campus per semester for brand-ambassador programs" [33]. While organizations like the Linux Foundation donate resources for open source advocacy, these resources are not at the levels required to compete with these campaigns.

The Linux community makes its own small, and at times very effective, efforts that have a great deal to offer international education. In a recent article on Linux.com, "The Unsung Heroes of Linux, Part I" readers are left with a message that "There is a moral to this story, and that is that Linux is more than giant wealthy companies, or glamorous celebrity geeks, or an unruly rabble...It is fundamental building blocks that anyone can learn to use to make the world a little bit better" [34]. Those advocating for Linux include, of course, individuals like Linus Torvalds, Eric Raymond, Mark Shuttleworth, and Richard Stallman.

4.4.2 Considering Contexts

While Stallman is recognized as "equal parts famous and infamous," his methods are effective—particularly in relation to educational settings [34]. Six years ago when Kerala, India, was considering migrating to Linux, a decision that effected 1.5 million students, "free software guru Richard Stallman [became] virtually the consultant to the Kerala government's IT initiatives" [21]. To help sway the government officials, Stallman traveled to India and "gave a presentation as to how free software has been an exciting education and computing model in a Spanish province" [21]. That "province" was no doubt the autonomous region of Extremadura, where Stallman had essentially served in a consultant capacity for their Linux development. Indeed, when Spanish scholars José Louis González Sánchez and Alfonso Gazo Cervero published their edited collection, *Sociedad y Software Libre: Curso Internacional de Verano de la Uex sobre Software (Society and Free Software: Summer International Conference of the University of Extremadura about Software)*, the first four chapters of the book are Spanish translations of Stallman's famous philosophical approach to and definitions of free software [35].

Stallman's willingness to work with these two governments to localize the otherwise global efforts of the (global) Linux community with a focus on education designed results. In other words, Maher hits the nail on the head when she highlights the importance of understanding open source projects in the context of "nationally-situated cultural values" [31, p. 367]. The case studies she points to as parts of her research (e.g., the Uganda Martyrs University case study written by Victor van Reijswoud and Emmanuel Mulo) are what technical communicators can be reading in order to understand more completely the how and the why of Linux success on the desktop. And, as important, educators in technical communication need to begin familiarizing themselves with open source products so they can teach students in order to better prepare them for working in global contexts. In the near if not immediate future, a global context will no doubt include the need to work and communicate with the millions of international Linux users.

In the next section, I will advocate taking our engagement with these case studies a step further by proposing that we work with our students to research and write our own case studies involving Linux on the desktop. I will accomplish this by profiling Extremadura region of Spain and their LinEx project. In so doing, I will also advocate downloading, installing, and using different Linux distributions as an effective means of studying and recognizing how and why the glocal works for Linux.

4.5 Extremadura

In the summer of 2007, I visited the Universidad de Extremadura in western Spain to meet with Professor José Louis González Sánchez in order to learn more about his university's reliance on Linux. As I discovered, it isn't just the university, but Extremadura's entire educational system and many of its government agencies that have adopted and are using Linux for both everyday and specialized computing activities.

4.5.1 The Regional Decision

The choice of open source was essentially the region's only option as the cost of the proprietary software required to meet the government's overall needs was projected to exceed 20 million euros (just over $25 million USD at the time) [36]. Extremadura, one of Spain's 17 autonomous regions, is also the country's poorest. Situated in the western portion of Spain, the region forms a border with Portugal. Of its three major cities Badajoz, Cáceres, and Merida, only Badajoz (136,319 inhabitants) has a population greater than 100,000 [35, p. 24]. The region's historic reliance on agriculture and its relatively low population density combined with very little IT infrastructure positioned Extremadura to be left out of the emerging information economy. Instead, Extremadura overcame the odds by developing its own localized, or what I will refer to as a "glocalized," version of Linux: a product dubbed LinEx. Interestingly, the economic conditions affecting software adoption choices in this region are by no means unique. Rather, they reflect a growing global trend—particularly among emerging economies—to turn to open source options in order to enter the global marketplace. Thus, an understanding of the dynamics involving software adoption and use in Extremadura can provide technical communicators with an example of how a significant and growing segment of the world's population selects and uses technologies.

4.5.2 Examining the Context

Professor Sánchez, my contact for examining this development, teaches at the University's campus in Cáceres—approximately a 4-hour train ride from Madrid. As the country's poorest region, Extremadura and Cáceres were a stark contrast to Madrid's metropolitan feel. But, thanks to the LinEx project, they aren't falling behind on technology and technology-enabled education. When I met Sánchez, he designed a new Sony Vaio laptop, the kind of computer you would expect to see running the latest Windows OS in the United States. His machine, like almost all of those found within the Extremadura public administration and educational systems, instead runs the LinEx operating system. The LinEx installation comes pre-loaded with OpenOffice® and Firefox®. It also has graphics programs like GIMP® and the OpenShot® video editor.

After showing me around the LinEx desktop, Sánchez presented me with a copy of his co-authored collection, *Sociedad y Software Libre*. We then discussed the ongoing development of the operating system. He also detailed meeting Stallman, who Sánchez described as "passionate" and, with a smile, called him a "character." Regardless, it was clear that Sánchez was grateful for Stallman's involvement with the LinEx project and quite pleased with the outcome.

It may be tempting to classify what the Extremadura region has done with its LinEx project as a mere translation of the Debian Linux distribution. I would argue that after speaking with Sánchez as well as reading interview responses from one of the LinEx lead developers, Daniel Campos, the project is much more complex [36]. Technically, Campos has had to contend with updating a standard installer to make it work with LinEx, creating a simplified control panel for easier use, and creating specific tools for managing LinEx in a classroom setting. While the LinEx developers did adopt a standard (and already available) Spanish translation for their operating system, Campos has found that they have needed to do some regionally specific localizing. The desktop has been rebranded for the region as well. The typical Linux penguin, "Tux" is not the LinEx icon. Instead, LinEx has a cartoon version on a stork, a bird common in the Extremadura region. Indeed, LinEx is not a translation, but a true localization or *glocalization*.

4.6 The Glocal

The amount of mainstream and scholarly writing dedicated to definitions, theories, and consequences of globalization is vast. Perhaps most famously, Thomas Friedman brought a particular definition of the term into popularity with his publication of *The World is Flat* [37]. Criticisms of that project are also not in short supply. In 2006, William Marling published *How 'American' is Globalization?* in which he wrote:

> Thomas Friedman has called this a "flat world," but that is wrong. The metaphor is anachronistic. There is nothing flat about a world in which 60 percent of China is not connected to international transportation and the United States ranks fourteenth in Internet broadband connections. Rather we are approaching a bilevel world – multivocal and confluently global at the same time. [38, p. 202]

For Marling, the global is always limited or reined in, at least in some potent capacity, by the local. That is, even in a globally designed and distributed open source project, some aspects of the local will persist. The cover of Marling's book, for example, features a photo that the author took of a little girl posing with the Japanese version of Colonel Sanders at a local KFC. People from the United States will recognize the colonel's iconic face, but might be surprised to see him dressed as a samurai warrior. Imports are adapted to local tastes and to meet the needs and the demands of local people and their culture.

Predating these debates by more than 15 years, sociologists have been developing ideas similar to Marling's "bilevel" dynamic with their own theories, including the coining of the term "glocal." In 1992, Roland Robertson explained that: "glocalization registers the 'real world' endeavors of individuals and social groups to ground or to recontextualize global phenomena or macroscopic processes with respect to local cultures" [3, p. 173-4]. It is my claim that the LinEx project exemplifies this definition of the glocal as it has taken Linux, a true "global phenomenon," and grounded it in the local Extremadura culture.

4.7 Situating Professional Communication Students in the Glocal

The professional communication workplace continues to become more international and successful projects of all kinds rely on the distributed and collaborative efforts of diverse groups of people. Pedagogically, it makes sense to situate our students amidst these changing spaces. In other words, the conclusion of this chapter places special emphasis on putting students into the position where they get a view of and get their hands on the glocal. Open source projects, specifically desktop Linux operating systems, provide excellent training for the modern global business context.

One of the simplest approaches to the glocal is to ask students to research their own Linux success story. As mentioned, many of the larger Linux-based companies like Red Hat and Canonical write and post case studies. (Copies of Ubuntu studies, for example, can be accessed at http://www.ubuntu.com/business/case-studies.) These studies often follow a conventional, short-form version of the case study genre and feature sub-sections dedicated to summary, challenge, solutions, and results. Getting a sense of the expectations of the case study genre sets the stage for professional communication students to begin researching and then writing their own Linux case studies. To do so, students can locate leads in blog posts, news articles, conference proceedings related to education and technology, as well as many of the popular sites dedicated to the Linux community. Students can then present and compare where they are finding these Linux success stories and begin to form hypotheses around the how and why of those successes.

In many respects, such an assignment isn't new (perhaps only new as its subject matter relates specifically to Linux). For example, using grounded theory and her own application of the glocal, Xiaoli Li offers a useful study on the cultural factors involved with resume writing in China versus the United States. She concludes her article by reflecting on the teaching potential related to her findings:

> To test students' understanding of analyzing the international audience, we may design our typical job application assignments to require students to apply for an international job or internship. In this assignment, students are required to demonstrate their understanding and their skills of how to apply what they have found about their audience to a particular situation instead of using any stereotyped information about people from a country or culture. [39, p. 277]

Similarly, studying successful (and unsuccessful) Linux distributions promotes the critical literacy Selber calls for, by essentially having students "seek oppositional discourses that defamiliarize commonsensical impressions of technology" [30, p. 88]. In other words, if students (or we as instructors) have reached a level of familiarity with Microsoft products that they are used without any conscious critique, it is up to instructors to create assignments that jar students from their technological comfort zones. Researching and writing case studies related to Linux on the desktop provides learning opportunities for students that inspire critical reflection and better prepares them for the global technical communication environment.

4.8 Using Linux on the Desktop

I think most professional communicators will agree that there is a significant difference between researching and writing about a topic, such as Linux desktop success cases, and downloading and using Linux on the desktop. I argue here that taking this important step of learning how to use the technology is essential to situating students in a glocal setting outside of their comfort zones.

There are a variety of methods available to help technical communication students run a version of Linux on their desktop (or to work with computer lab administrators to run Linux). Downloading a Linux distribution (one can download such programs from Ubuntu at: http://www.ubuntu.com/download) and burning it to a CD/DVD or saving it to a jump drive offers the luxury of booting Linux directly from either medium. Or, built-in Linux installers (e.g., Ubuntu's Windows installer for the desktop: http://www.ubuntu.com/download/desktop/windows-installer) assist users with the process of partitioning, for example, the hard drive on their Windows machine so they may make room for Linux. At startup, users may choose to then boot in Windows or Linux. Such activities constitute a true first step in developing the kind of critical literacy students will need to better assess different technologies and consider what technological options might be best to use to communicate in different global and local contexts.

Again, according to Selber, another key component to a student's developing critical literacy is the student's control over his or her own computing experience. That experience should include control over the hardware and software students use to communicate/share ideas with others. The following discussion provides details on my own efforts to prepare assignments that involve downloading and using Linux in a technical communication classroom.

In an effort to empower (and awaken) students, I first ask them to download and install Oracle VM Virtual Box® (virtualbox.org). At this stage of their professional communication education, a majority of the students in class do not know about emulation and virtual machines. The idea that a piece of software can be installed on their computer and run on their existing operating system with the sole purpose of serving as a "host" with which to run other operating systems usually comes as a surprise. The Oracle Virtual Box fits well with this assignment because the software itself is open source and licensed under the GPL. Also, it is a great tool for collecting, organizing, running, and comparing alternate operating systems.

The setup related to this initial process is easy; the only major decision a user has to make is how much RAM should be allocated to the virtual machine (and alternate OS) while it is running. In my experience, most Linux distributions don't require more than 1 GB of RAM allocated to the virtual machine in order to run effectively. The final added bonus of Oracle Virtual Box is that the user can switch between the host operating system (for example a version of Windows or a Mac OS) and the virtual operating system they are experimenting with. This feature makes points of comparison easier to see and document.

Figure 4.1 displays a screen capture of the Oracle Virtual Box Manager and the left panel of the manager displays the alternate operating systems I may run on my host OS. This screen capture shows that I may run either a 32-bit version of Windows 7 or LinEx inside of Virtual Box. In this case, I have installed Oracle Virtual Box on a Macbook Air with 4 GB of RAM. With this current configuration, I am able to demonstrate to students the unique affordances of all three operating systems without having to reboot the machine; I can run both Windows 7 and LinEx in the virtual machine.

Clicking "Start" in the Manager will "boot" the selected operating system, and that system will then run inside of the virtual machine. I ask students to use and critique their new Linux distributions and probe them for their different features. I also ask students if they can find evidence of their distribution demonstrating a particular local trait that might lead us into discussions on Linux and the glocal. The educational objective of these activities is to empower students to explore other desktop operating systems by

USING LINUX ON THE DESKTOP 85

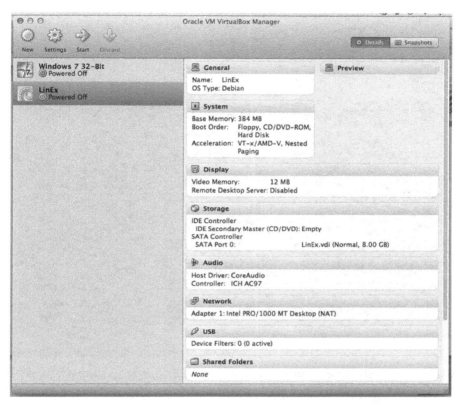

FIGURE 4.1. Oracle VM Virtual Box Manager. The Oracle Virtual Box Manager displays a list of operating systems available to run on the Virtual Box platform as well as information about each OS.

mastering Oracle's Virtual Box. That is, students understand how to functionally use Virtual Box as a tool but also grasp the concept of virtualization on the desktop.

From this point, students are then asked to identify particular local traits that may be found in a Linux distribution like LinEx. What students are looking for might be something as obvious as the rebranded look of LinEx that features their Extremadura stork (see Figure 4.2) or might be more subtle like the curious omission of a particular application. Why might, for example, a particular distribution be released without any games? If students' research shows that a particular distribution is going strong, what can we learn from using it that reveals more about the how and the why of that success? Conversely, what might students conclude if the Linux distribution they are researching displays no clear localized traits? In short, this tiered assignment that asks students for research into existing case studies, for the research and writing of their own case study or report, and for a critical engagement with an actual "glocalized" Linux distribution should help prepare future technical communicators for our shifting global, local, or

FIGURE 4.2. Login screen for LinEx. Once the LinEx operating system is launched within Oracle's Virtual Box, users are greeted with a login screen to begin using the OS.

glocal landscapes. Again, such activities help students prepare for a global technical communication landscape where their partners, customers, and colleagues understand and experience desktop computing differently.

It should be noted that in order to more efficiently assess (and initially implement) a Linux component such as this assignment sequence into an existing course, instructors might wish to break the class into small teams of approximately three students each. The collaborative effort of a group project can yield a more thorough case study as well as help alleviate any anxiety students may have with downloading and exploring a new open source operating system. As mentioned, case studies representing Linux are typically brief reports of one to three pages and feature sub-sections dedicated to summary, challenge, solutions, and results. Students and teachers can use existing case studies as models and rubrics as an introduction to and assessment mechanism for this genre. While all of the sub-sections of a case study are important, the solution section will document how Linux was employed and perhaps most importantly, how it was customized to meet a local challenge.

Next, students may be asked to demonstrate proficiency with Oracle Virtual Box as they use it to successfully run their Linux distribution. In order for me to assess my students, it was, as noted, more efficient to break the class into small teams. Student teams would come to the front of the class and present their findings from their case studies as well as "demo" their Linux distribution. As a result, students demonstrate publicly their abilities to master the functional aspects of using a tool like Oracle Virtual Box as well as their critiques and analyses of alternate desktop computing environments. These open source environments may already be playing a larger role in the new global technical communication landscape where students will meet their future partners, customers, and colleagues.

4.9 Conclusion

One cannot be an effective digital writer without knowing both technical procedures and how to deploy them to achieve the desired end. [39, p. 207]

The irony or perhaps the tragedy here is that during the course of researching and writing this chapter the LinEx distribution developed for Spain's Extremadura region has ceased to exist. This is not a willful abandonment on the part of the project team or the result of documented technical or functional inadequacies with the software; the reasons for LinEx's demise are entirely political. At the end of 2011, the PSOE or the Spanish Socialist Worker's Party lost control as the country's ruling political party and was supplanted by the more conservative PP or People's Party. In an effort to streamline any government spending, the new ruling party required that computing initiatives receiving government funding be turned over to a centralized agency in Madrid [40]. That is, localized initiatives like LinEx that were designed to meet the specific needs of autonomous regions like Extremadura would go away.

On an open forum dedicated to discussing the Debian Linux distribution (again, Debian is the version of Linux on which LinEx was based), developers associated with LinEx expressed their frustrations with pulling the plug on LinEx as well as their theories regarding the role politics played in the decision. Lead developer José Luis Redrejo Rodríguez wrote:

> The new people in charge of the Extremadura government don't like the good press and name that Linex, and the free software, gave to the previous party in the government. And they want to change things. I don't say they're going to remove all the free software we have in education (I don't think that's technically possible, and also we can not afford it), but they maybe will move from Debian to Ubuntu or to OpenSuse or Fedora. They are firing all the people who made the previous situation possible. [41]

As if Linux on the desktop doesn't have enough challenges ahead of it, in the case of LinEx, it appears to have been the victim of its own success. The bright spot, if there is one in this situation, is that the new ruling government party appears to be interested in staying with open source software. Whether or not the decision is made to use a more popular Linux distribution like Debian, Ubuntu, or Fedora remains to be seen. Whatever the choice, preparing future technical communicators to work in a global economy means preparing them to work with open source. Fortunately, that preparation is much easier now than in Linux's past.

Linux is now 20 years old, and in its earlier days, its use required a great deal of determination, dedication, and technical savvy on the part of the user. Today, proponents of Linux like to trumpet that such levels of engagement are no longer required; one can download, install, and run Linux in a matter of minutes. Such ease of use is certainly a triumph for the Linux community and a testament to its thousands of volunteer developers. That same ease of use, however, has not pushed the forecasts for Linux's market share on the desktop any higher. Indeed, the how and the why of the small percentage of desktop successes may be linked back to the ability to customize and tinker, enabling a particular Linux distribution may make sense glocally.

In an effort to capture the spirit of customization, Neal Stephenson famously drew an analogy between participation in the hacker community and the painstaking ownership of an old MGB sports car. Apparently, a friend's dad had a commitment to the upkeep of the vehicle despite its many faults:

> Sure, the MGB was a lousy car in almost every way that counted: bulky, unreliable, underpowered. But it was fun to drive. It was responsive. Every pebble on the road was felt in the bones, every nuance in the pavement transmitted instantly to the driver's hands. He could listen to the engine and tell what was wrong with it. The steering responded immediately to the commands of his hands. To us passengers it was a pointless experience in going nowhere…But to the driver it was an experience. [42, p. 5]

Any successful attempts to grow Linux numbers on the desktop will benefit from inspiring an attention to the glocal that, in turn, I believe, will result in the overall experience that Stephenson describes. Any successful attempts by educators to introduce students to the "competing discourse" of Linux will benefit from some level of hands-on engagement with open source technologies. As technical communication instructors, it is our job to provide students with the critical faculties to recognize the benefits of working in, around, and with technologies that at once possess the fortitude to survive in competitive global spaces but are also be sensitive to local conditions and needs. In other words, we must pay attention to the glocal.

References

1. Gartner Says Windows 7 Will Be Running on 42 Percent of PCs in Use Worldwide by the End of 2011. Available at: http://www.gartner.com/it/page.jsp?id=1762614.
2. Operating Systems Desktop Share. Available at: http://www.netmarketshare.com/.
3. R. Robertson, *Globalization: Social Theory and Global Culture*, Thousand Oaks, CA: SAGE, 1992.
4. The 'Year of the Linux Desktop' Isn't Coming | ZDNet. Available at: http://www.zdnet.com/blog/hardware/the-year-of-the-linux-desktop-isnt-coming/16022.

REFERENCES

5. The Story of Linux: Commemorating 20 Years of the Linux Operating System. Available at: https://blogs.oracle.com/linux/entry/the_story_of_linux_commemorating.
6. Open Technology within DoD, Intel Systems. Available at: http://archive09.linux.com/feed/61302.
7. Gartner Says Sales of Mobile Devices Grew 5.6 Percent in Third Quarter of 2011: Smartphone Sales Increased 42 Percent. Available at: http://www.gartner.com/it/page.jsp?id=1848514.
8. Intrinsyc Joins Open Handset Alliance. Available at: http://www.marketwired.com/press-release/intrinsyc-joins-open-handset-alliance-tsx-ics-1521252.htm.
9. Google Announces $10 Million Android Developer Challenge. Available at: http://googlepress.blogspot.com/2007/11/google-announces-10-million-android_12.html.
10. How Much Money Does Google Make from Mobile?. Available at: http://www.zdnet.com/article/how-much-money-does-google-make-from-mobile.
11. S. Weber, *The Success of Open Source*, Cambridge, MA: Harvard University Press, 2005.
12. M. Fink, *The Business and Economics of Linux and Open Source*, Upper Saddle River, NJ: Prentice Hall, 2002.
13. P. Himanen, *The Hacker Ethic*, New York: Random House Trade Paperbacks, 2002.
14. K. St.Amant, *Handbook of Research on Open Source Software: Technological, Economic, and Social Perspectives*, Hershey, PA: IGI Global, 2007.
15. C. M. Kelty, *Two Bits: The Cultural Significance of Free Software*, Durham, NC: Duke University Press, 2008.
16. E. S. Raymond, *The Cathedral & the Bazaar: Musings on Linux and Open Source by an Accidental Revolutionary*, Cambridge, MA: O'Reilly Media, 2001.
17. Desktop for Business. Available at: http://www.ubuntu.com/business/desktop.
18. Andalusia Deploys 220,000 Ubuntu Desktops in Schools Throughout the Region. Available at: https://insights.ubuntu.com/2010/03/13/andalusia-deploys-220000-ubuntu-desktops-in-schools-throughout-the-region/.
19. Linux for the Long Haul. Available at: http://www.linuxjournal.com/magazine/linux-long-haul.
20. Russian Schools Move to Linux. Available at: http://news.bbc.co.uk/2/hi/technology/7034828.stm.
21. Kerala Logs Microsoft Out of Schools. Available at: http://www.rediff.com/money/2006/sep/02microsoft.htm.
22. Korea Brings Homegrown Open Source to Schools. Available at: http://news.cnet.com/Korea-brings-homegrown-open-source-to-schools/2100-7344_3-5755892.html?tag=mncol;txt.
23. Robert Litt on Building a Computer Lab for Free at the EdSurge DIY Learning Pavilion. Available at: https://showyou.com/teach4america/y-nZ84GcDGoMw/robert-litt-on-building-a-computer-lab-for-free.
24. How One Teacher Built a Computer Lab for Free. Available at: http://ifixit.org/blog/3001/how-one-teacher-built-a-computer-lab-for-free/.
25. Linux-based Desktops, Servers, and Curriculum at Private Christian School. Available at: https://schoolforge.net/education-case-study/linux-based-desktops-servers-and-curriculum-private-christian-school.
26. Open Source at the National Education Computing Conference. Available at: http://arstechnica.com/old/content/2006/07/7361.ars.
27. After Xandros, then what? Available at: http://1topix.com/forums-xandros-com-index/.

28. FTC Charges Indiana Firm with Failing to Live Up to Rebate Promises. Available at: http://www.ftc.gov/opa/2009/01/mds.shtm.
29. C. Smith, Personal interview. February 21, 2012.
30. S. Selber, *Multiliteracies for a Digital Age*, Carbondale, IL: Southern Illinois University Press, 2004.
31. J. H. Maher, "The technical communicator as evangelist: Toward critical and rhetorical literacies of software documentation," *Journal of Technical Writing and Communication*, vol. 41, no. 4, pp. 367–401, 2011.
32. Effective Evangelism. Available at: http://zgp.org/~dmarti/linuxmanship/Comes-3096.pdf.
33. On Campus, It's One Big Commercial. Available at: http://www.nytimes.com/2011/09/11/business/at-colleges-the-marketers-are-everywhere.html?_r=0.
34. Unsung Heroes of Linux, Part One. Available at: https://www.linux.com/learn/tutorials/543893-unsung-heroes-of-linux-part-one.
35. J. Louis González Sanchez and A. Gazo Cervero, *Sociedad y Software Libre: Curso Internacional de Verano de la Uex sobre Software*, Badajoz, Spain: Indugrafic, 2004.
36. gnuLinEx 2004 Launched. Available at: http://www.linuxjournal.com/article/7908.
37. T. L. Friedman, *The World Is Flat 3.0: A Brief History of the Twenty-first Century*, New York: Picador, 2007.
38. W. H. Marling, *How "American" Is Globalization?*, Baltimore, MD: The Johns Hopkins University Press, 2006.
39. X. Li, "A genre in the making—A grounded theory explanation of the cultural factors in current resume writing in China," *IEEE Transactions on Professional Communication*, vol. 54, no. 3, pp. 263–278, 2011.
40. Extremadura Abandona Linex. Available at: http://www.publico.es/ciencias/extremadura-abandona-linex.html.
41. Spanish's Extremadura Abandoning Linex, but Maybe Adopting Debian. Available at: https://lists.debian.org/debian-project/2012/01/msg00007.html.
42. N. Stephenson, *In the Beginning…Was the Command Line*, New York: Avon Books, 1999.

5

Teaching the Ethics of Intercultural Communication

Dan Voss
Lockheed Martin Corporation

Bethany Aguad
University of Central Florida

Communicating across cultures is challenging. Resolving ethical issues across cultures is at least as challenging. It therefore follows that the ethics of intercultural communication would be more challenging still. The issues are complex. Effective communication is rooted in the fundamentals of rhetoric; resolving ethical issues boils down to understanding values and doing what is right—in cases of ethical conflicts, defining "right" as that which serves the highest, or most core, value. What if different cultures define ethical values differently? Can what is considered "wrong" in one culture be "right" in another? This chapter attempts to blaze a safe trail through that thorny thicket—one that helps teachers and trainers provide useful instruction that retains flexibility so as not to become a formulaic process, prey to the dangers of stereotyping and ethnocentrism.

Teaching and Training for Global Engineering: Perspectives on Culture and Professional Communication Practices, First Edition. Edited by Kirk St.Amant and Madelyn Flammia.
© 2016 The Institute of Electrical and Electronics Engineers, Inc. Published 2016 by John Wiley & Sons, Inc.

5.1 Introduction: Globalization Introduces an Intercultural Dimension to Business Ethics

The phenomenon of globalization has led to increasing emphasis on teaching intercultural communication, both within the curriculum of engineering degree programs and via in-service training of practicing engineers. Teaching the ethics of intercultural communication is an area that has not heretofore received much attention, either in academe or in industry. The ethical conflicts likely to arise while doing business across international and cultural boundaries, however, are quite significant and warrant a closer look in the teaching/training of the technical community. Teaching the ethics of intercultural communication is especially challenging, because it represents the intersection of teaching ethics and teaching intercultural communication, both of which must be separately addressed if we are to do justice to the much narrower overall topic. Fortunately, we may safely assume that teaching intercultural communication is receiving ample treatment—as in every other chapter in this book. But before attempting to teach the ethics of intercultural communication, it is fitting for us to look at teaching ethics in general, to establish a framework. Doing so will enable us to focus on the narrower—yet nonetheless important—topic of this chapter.

5.1.1 Value Analysis Superimposes Cultural "Filters" on Classical Ethical and Contemporary Value Models

A look at relevant literature shows much has been done on teaching ethics in a technical context and on teaching intercultural communication in a technical context. The body of work grows much sparser, however, when one tries to "zero in" on the intersection of the two. We have decided to begin by noting core readings, starting with a brief survey of classical ethical models, proceeding to describe two baseline value models typical of US/Western European culture, and then presenting the ethical codes and guidelines of two technical companies and four professional organizations of engineers and technical communicators. That accomplished, we can then superimpose additional "filters" to consider the differing values of divergent international cultures, ethnicities, religions, and other ethical frames of reference. Just as solving an equation with multiple unknowns requires multiple steps, exploring the topic of ethics across cultures demands multiple analytical perspectives.

To impose order on this complex topic, we explain how to use value analysis as a "scientific" approach to identify and resolve ethical conflicts using the two baseline value models and representative ethical codes within the technical community. Next, we introduce the intercultural "filter," which complicates the equation by bringing into play value systems that, in some cases, are quite different from those of Western culture. We then explore the critical concept of cultural/moral relativism versus cultural/moral universalism. The process flow diagram in Section 5.7 encapsulates the process of value analysis in the specialized context of this chapter, superimposing both ethical value models and cultural filters to identify and resolve ethical conflicts in intercultural technical communication.

5.1.2 Teaching the Ethics of Intercultural Communication Applies to Technical Students and Professionals

The heart of the chapter is, naturally, about teaching the topic. We therefore address this focus by examining it from three contexts:

- For academic instructors teaching intercultural communication to engineering or science students
- For professors of engineering or science courses who would like to include intercultural communication within their curriculum
- For industry trainers teaching intercultural communication to the technical community

The emphasis of this examination is not on "right" versus "wrong"—which would oversimplify a complex subject. Rather, we focus on defining alternative courses of action and predicting the outcomes likely to result from such actions.

5.2 Literature Review Represents the Intersection of Ethics, Intercultural Communication, and Science/Engineering

In this age of global citizenship, students as well as professionals need to be prepared to face the challenges of responsibly and ethically navigating cultural differences. Professional engineers and scientists as well as students of engineering and science need to be prepared to address the differences in values between their personal beliefs, their professional values, and differences in ethical standards between cultures. How can anyone define an overarching code governing ethical behavior that is flexible enough to adapt to diverse cultural values?

5.2.1 Defining Ethics Becomes More Complex in an Intercultural Context

Determining what will benefit humanity is not always a simple matter of common sense. In a review of different journal articles on the subject of an engineer's social responsibility, Zandvoort concluded that attention must be given to developing an educational program that encourages engineering students to evaluate the organizational, social, legal, and political context of their work [1]. He believed that only after carefully considering these factors could an engineer make a socially responsible decision. Zandvoort suggests that current engineering courses focus too much on the ethical choices of the individual and not enough on the responsibilities of the profession as a whole.

Education in ethics alone, however, is not sufficient. It is important to integrate an intercultural perspective to render the ethical values and principles relevant in crosscultural contexts. Riemer recognizes the need to educate engineering students in intercultural communication because of economic globalization. He refers to Riordan's assertion that "developing a good cultural sensibility is not only good business, it is the ethical

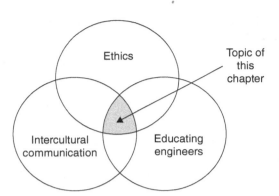

FIGURE 5.1. Research on the ethics of intercultural communication. Little research and few resources exist on teaching the ethics of intercultural communication to a technical audience.

thing to do" [2, p. 197]. Riemer suggests that gaining an understanding of cultural considerations and value dimensions would prepare students for communicating efficiently and effectively across cultures.

As depicted in Figure 5.1, this chapter focuses on the intersection of three different topics. Currently, limited research on teaching the ethics of intercultural communication to a technical audience is available. When interviewed, Dr. Paul Dombrowski attributed this to the "abstruse nature of ethical theories." He suggested that the education of technical audiences generally has an "emphasis on hard facts," which is difficult to incorporate into an ethics course with its inherently intangible principles [3]. Add the dimension of intercultural differences, and you are dealing with a very complex issue, Dombrowski observed. Yet this situation does not diminish the importance of education on the subject.

The previous research on this subject only considers specific aspects of teaching intercultural ethics to engineers. Luegenbiehl, for example, discusses the globalization of engineering ethics as it related to the principle of autonomy [4]. In so doing, he examines the roles and responsibilities of the engineer with regard to ethical differences in the United States and Japan. In the United States, for example, the burden of responsibility typically falls on the individual or on the profession as whole, whereas in Japan, the corporation serves to meet the needs of society. For example, when US engineers were asked what they did, they generally responded by noting they were engineers, and thus identify themselves by their profession. When presented with the same question, many Japanese engineers replied by noting the company they worked for (e.g., "I work for Mitsubishi" versus "I work as an engineer").

Based on these findings, Luegenbiehl proposes that a universal code of engineering ethics would have to account for cultural differences. He leaves room open to explore how to develop such a universal code and, once developed, to determine how effectively it could be implemented. (Later in this chapter, we discuss the difference between *universalism* and *cultural relativism* and describe how these perspectives influence ethical decisions. Rather than presenting one view as correct, we provide an alternative view—*contextual relativism*.)

Radebaugh, Soschinske, Rimmington, and Alagic mirror this sentiment and address the need for a new tool for teaching the subject, stating that the "challenge of teaching engineering ethics in a multi-cultural environment must include an approach to effective cross-cultural communication" [5, p. 4]. They proposed a new interactive approach to ethical simulations called "cage-painting" based on resolving cultural misconceptions to make the best ethical decisions. The *cage* is our construction of the world that filters all communication. In *painting* their cage, people reflect on their own personal and cultural perspectives. The goal is to paint their colleague's cage by gaining an understanding of her/his perspective through open dialogue.

5.2.2 Teaching Ethics Demands a Systematic, Analytical Approach to a Subjective Topic

While both Luegenbiehl and Radebaugh et al. provide useful information on the topic, neither provides a systematic approach for identifying and resolving ethical dilemmas in intercultural communication. In this chapter, we offer value analysis as a strategy for analyzing ethical conflicts and determining appropriate action. Integrating value models and different cultural values helps bring into sharper focus all the factors that influence the ethical decisions scientists and engineers must make in this ever-more-global economy.

Due to the recognition of the vital role the engineering profession plays in the greater good, many sources on educating engineers in ethics are available. Less research is available on teaching the ethics of intercultural communication in general and on teaching intercultural communication to the technical community in particular.

Newberry addressed some of the fundamental issues with ethics education within the technical field [6]. He considered the dangers of teaching ethics only on a superficial level, which, he asserted, could lead to stereotyping. Newberry's three main objectives for engineering educators were to encourage students to want to make ethical choices, to know how to make the right ethical decisions, and to understand the current guidelines for ethical practice of their profession.

At a time when failures in engineering ethics are prominently displayed through media, strong classroom or training emphasis is often placed on teaching *preventative ethics*, which requires examining past ethical failures and using the knowledge gained to prevent future engineering disasters [7]. However, Harris proposes an alternative to this common practice—*virtue ethics* [8]. A virtuous engineer consistently exhibits good behavior in accordance with her/his personal sensibilities. Harris suggested applying virtue ethics to engineering by achieving technical excellence, having an awareness of the way technology affects society, respecting the environment, and committing to the public good. Harris refers to Aristotle's explanation of good and virtuous behavior, which is discussed in the classical ethics section of this chapter.

In accord with Harris, Ramani made the case for a greater understanding of ethics through the pursuit of personal ethics. Ramani summarizes personal ethics as individuals doing what is right when no one is looking, and he calls out others who behave unethically [9]. The codes of ethics of professional societies or corporations provide some guidance. Consistent ethical behavior through a technical career can, however, only be achieved through dedication to lifelong learning, beginning in the academic classroom and then reinforced by ongoing industry in-service training and education.

The reality is that ethics always matter in the workplace, but the public eye only sees the failures. Dombrowski warns against ethics being "technologized," meaning reducing ethics to a technology of sorts, composed of procedures [10]. In his exploration of the space shuttle *Challenger* disaster, Dombrowski points out that instituting a defined procedure at an organizational level could not take the place of personal ethical responsibility. While ethical standards have their place, doing what is right ultimately comes down to personal judgment and responsibility.

5.2.3 Teaching Ethics within the Science/Engineering Community Raises Sensitivity to the Complexity of the Subject

"Doing the right thing" might sound simple, but numerous pressures come to bear on an individual when he or she is confronted with an ethical dilemma. Geistauts, Baker, and Eschenbach attempt to illustrate the complexity of pressures impinging upon an individual engineer's ethical decisions via a system dynamics approach [11]. They warn against perceiving the values of the individual, the organization, and the profession as static, because these values are all constantly evolving in response to the environment. Geistauts, Baker, and Eschenbach use causal loop diagrams to examine the linkages between factors influencing ethical decisions. These factors include the values of the profession, personal beliefs, obligations to the organization, and—unfortunately, but realistically, in most cases—also budget and schedule pressures.

This idea brings up the concern that with all the complex factors involved in an ethical decision, it might not be possible to teach ethics effectively to the technical community, let alone across cultures. Addressing this issue, Hashemian and Loui evaluate the success of instruction in engineering ethics at prompting feelings of professional responsibility [12]. In interviews of engineering students at the University of Illinois, they found that students who had taken *Engineering Ethics* as an elective were able to consider actions and outcomes and act more confidently and consistently than students who did not take the course. Furthermore, students who had taken the course expressed a greater inclination to intercede in situations where unethical behavior was occurring even when it was not their direct responsibility. Although the interviews are not conclusive proof that education leads to more ethical behavior, this study does indicate that students at least to some degree internalize the ethical principles they are taught.

5.2.4 Teaching Ethics for the International Technical Community Requires Awareness of Variations Across Cultures

While learning the principles of ethics is necessary, it is not sufficient to behave ethically in an intercultural context. Accounting for the differing value systems of divergent cultures requires an awareness of the principles of intercultural communication as well as ethics. The research on the ethics of intercultural communication is less abundant than it is on engineering ethics.

> While learning the principles of ethics is necessary, it is not sufficient to behave ethically in an intercultural context. Accounting for the differing value systems of divergent cultures requires an awareness of the principles of intercultural communication as well as ethics.

To achieve successful and ethical intercultural communication, Flammia and Voss call for an educated, in-depth awareness of cultural differences, beyond the superficial [13]. After establishing a framework by examining classical ethical models, they focus on five key values for successful intercultural communication: "legality, privacy, teamwork, social responsibility, and cultural sensitivity" [13, p. 73]. Flammia and Voss conclude that a sensitive intercultural awareness requires generalized knowledge of the political, social, religious, and economic values of other cultures while avoiding the pitfall of categorizing individuals based on stereotypes. They also recommend going beyond simple legality to achieve ethical behavior, citing Jim Crow laws as an example of how a mandate from an unprincipled governing agent can be "legal" but clearly not ethical.

Respecting the rights of the individual to privacy requires learning the different expectations for privacy between cultures. Appreciation of the differences in intercultural perspectives leads to successful teamwork in international partnerships. To fulfill the duties of social responsibility, every individual must constantly balance the ethical values of the profession with the realities of the business world—realities which, in turn, often differ across cultures and nations.

In a 2012 personal interview, Dr. Pavel Zemliansky expressed his belief that teaching intercultural communication to a technical audience is very important, yet he pointed out that research and resources on intercultural communication and engineering education are scarce [14]. In his experience with teaching across disciplines, Zemliansky has found the best way to learn intercultural communication is by experiencing it. While some bedrock principles, like not fabricating results in research, are universally accepted in the professional community, Zemliansky cautioned that engineers should never assume they are fully cognizant of the views of other cultures. Concepts that citizens from the United States might assume are universally understood, like "freedom," can have significantly different interpretations or expressions in other cultures. Students who are part of cross-cultural teams can thus benefit enormously from having an open dialogue, learning—and hopefully *un*learning—many misconceptions they have about other cultures.

5.2.5 Working on International Teams Hones Engineering Students' Intercultural Communication Skills

Zemliansky is not the only academician who favors this approach to teaching intercultural communication to the technical community. De Graaf and Ravesteijn express the need for a "complete engineer" who is able to communicate interculturally [15]. They point out that many universities fail to adequately prepare students to participate in intercultural teams. As a consequence, they contend, international companies bear the responsibility of filling the gaps in the education of their engineers. It appears to the authors that *both* the academy and industry have a responsibility in this area, and if both address that responsibility, the result will be increased intercultural awareness within the technical community.

To this end, Escudeiro, Escudeiro, Barata, and Lobo propose having engineering students work on international teams to solve engineering problems [16]. Such experiences would give students the intercultural skills they need to work successfully in the twenty-first century engineering environment across national, ethnic, and cultural lines. In their Multinational Undergraduate Team Work (MUTW) project, Escudeiro, Escudeiro, Barata, and Lobo have students from different countries work together to create a joint capstone project. In so doing, they recognized Dombrowski's idea that:

> To enact thoughtfully our ethical responsibility as individuals, we need to understand what others have thought on the subject. We also need to understand what those who are affected by our decisions think and feel about ethical responsibility. We learn this by studying and communicating actively with those around us. [17, p. 5]

Based on positive feedback from students, Escudeiro et al. call for a greater focus on intercultural competencies in engineering education, and they offer experience on heterogeneous international teams as the best solution.

Ultimately, both ethical responsibility and cultural sensitivity fall upon each engineer or scientist, but proper education enhances the entire profession by teaching engineers to act ethically as global citizens. In *Understanding Technology,* the late Charles Susskind, professor emeritus of electrical engineering at the University of California, Berkeley, proposes a type of Hippocratic Oath for engineers that emphasized the personal responsibility of every engineer [18]. Engineers swearing to this oath promise, to the best of their ability, to undertake projects that would benefit humanity; not to allow personal preferences regarding religion, politics, or nationality to interfere in completion of work; and never to violate the "laws of humanity" [18]. This concept is further explored in the discussion of universalism versus cultural relativism later in this chapter.

5.3 Four Classical Ethical Models Form the Foundation for Studying the Ethics of Intercultural Communication

Building on the work of the researchers in intercultural communication and ethics cited in the review of literature, we hope to suggest ways in which professors of engineering and science as well as industry trainers who work with a technical community can teach the ethics of intercultural communication. Because our review of the literature identified few sources that focused on that narrow topic, we are taking the following approach to examine the topic in more detail:

- Surveying classical models of ethics
- Introducing value analysis, a systematic approach to identifying and resolving ethical conflicts

- Surveying six value models, showing how to use them with value analysis
- Applying value analysis in an intercultural context
- Discussing moral/cultural relativism versus moral/cultural universalism

Once we have completed these steps, we will "zero in" with specific suggestions on how to teach this narrow yet important topic, both in academe and in industry, and provide resources and tools to assist professors and industry trainers in so doing.

5.3.1 Aristotle, Confucius, Kant, and Mill Offer Four Fundamental Ethical Constructs

Before we can discuss how to resolve ethical dilemmas in intercultural situations, we must establish a "baseline" for ethics and values. Four classical models of ethics serve to establish a contextual framework. First, in *Nicomachean Ethics,* Aristotle proposed a code of ethics based on the habits of good character, rather than prescribing standards for behavior. Aristotle begins his inquiry into ethics by establishing the nature of "the good," which "has rightly been declared to be that at which all things aim" [19]. In pursuit of this aim, "human good turns out to be activity of soul in accordance with virtue" [20].

Aristotle took a practical viewpoint on ethics. To achieve the highest good, usually considered happiness, an individual must act virtuously. Aristotle believed that ethical virtues were moderate, straying neither to the side of excess nor deficiency. He believed it was the responsibility of the community to train each person to behave properly [19]. While Aristotle's balanced behavior promotes tolerance and his quest for the highest good stimulates a culturally sensitive approach to intercultural ethics, he does not provide any systematic procedure for making ethical decisions. Of particular concern is the fact that "the highest good" is likely to vary from culture to culture [20].

Confucianism is an ethical and philosophical system with a humanist perspective, and this system has influenced the development of ethics in many of Eastern cultures. Centuries after Aristotle, Confucius (commonly referred to as *Kong Zi* in modern China), proposed a similar approach to ethical behavior. According to Confucius, every individual should pursue virtue. Rather than learning rules or restrictions, followers of Confucius develop their judgment and cultivate sincerity in their actions. Primarily, an individual follows ritual propriety [21].

In the early Confucian writing, the concept of "lǐ" represents the rules for proper behavior [21]. These rules require individuals to take the appropriate action at the appropriate time in order to maintain a careful balance between respecting ritual and disregarding tradition to achieve a greater good. Following Confucian ethics, the roles people play in relation to those around them determine what behavior is proper [22]. Within this system of ethics, Confucius also taught a variation of the "Golden Rule": "What you do not wish upon yourself, extend not to others" [21]. Thus, Confucian ethics support a way of approaching intercultural differences, but they do not explain how to solve issues that arise from different people's perspectives.

Utilitarianism, championed by John Stuart Mill, stipulates that "it is the greatest happiness of the greatest number that is the measure of right and wrong" [23]. Mill believed that ethical actions contribute to the happiness of all people beyond the individual. He defined happiness as "pleasure, and the absence of pain" [24]. Following this consequentialist ethical model, the consequences of an action supersede any question of an immutable moral standard. While seemingly sensitive to ethical variations, this model can lead to the violation of the rights of the individual in favor of a perceived greater good.

Immanuel Kant offered a different perspective on rules for directing behavior [25]. Kant proposed a series of maxims that could be applied in many circumstances due to their abstract nature:

> You must first consider your actions in terms of their subjective principles, but you can know whether this principle holds objectively only in this way: that when your reason subjects it to the test of conceiving yourself as also giving a universal law through it, it qualifies for such a giving of universal law [25, p. 226].

Essentially, Kant's *categorical imperative* requires that all people make choices that conform to what they think should be universal laws. The personality and preferences of the individual, as well as her/his circumstances, should play no part in determining what actions should constitute universalities. Kant believed his code of ethical behavior should apply universally across cultures. (This idea ties in with our discussion of universalism versus cultural relativism that appears in the next section of this chapter.) The difficulty, obviously, is determining what values comprise the universal code.

Despite their major differences, all of these models of ethics seek to provide guidance in making the right choices. However, according to Dombrowski, learning the ethical models or standards is not enough to achieve consistent ethical behavior. In the end, it still comes down to the attitude of the individual [3].

5.3.2 The Path Between Universalism and Relativism Leads into a Thorny Ethical Thicket

Communicating with other cultures requires everyone to make ethical decisions when problems arise. Traditionally, two approaches have been used to address such situations involving intercultural ethics: universalism and relativism. Because neither approach, by itself, suffices in guiding individuals to make the right decisions, an ethical engineer or scientist must learn to reconcile these two divergent viewpoints.

Universalism is founded on the belief that certain ethical standards do exist that seem to apply to all cultures [26]. Kant declared that "codes of conduct and morality must be arrived at through reason and be universally applicable to all societal environments at all times" [27, p. 7]. This universal approach places the burden of moral and ethical responsibility on the individual regardless of his or her cultural surroundings.

Essentially, a universalist judges actions based on intent rather than outcome. In so doing, that person ignores cultural, sociological, religious, and ethnic variations when determining what behavior is "right." The difficulty with universalism arises from our inability to determine what the universal code for ethical behavior should be—to define what is right. The problem is that *no standard for ethical behavior is free from cultural context.*

> Universalism is founded on the belief that certain ethical standards do exist that seem to apply to all cultures. Relativism is based on the premise that it is not possible to develop one absolute code of ethics that applies uniformly across all cultures.

Relativism, in contrast, is based on the premise that it is not possible to develop one absolute code of ethics that applies uniformly across all cultures. Franz Boas, pioneer of anthropology, first suggested that "civilization is not something absolute, but ... is relative, and ... our ideas and conceptions are true only so far as our civilization goes" [28, p. 589]. Cultural relativism, in turn, holds that right behavior can only be determined within the bounds of a given culture [26]. This perspective acknowledges that cultural values shape ethics.

Both the universalist and the relativist perspectives become manifestly unreasonable if extended too far. For example, when taken to the extreme, moral absolutism dictates a man be sentenced to nearly 20 years in prison for stealing a loaf of bread to feed starving children, as in Victor Hugo's 1862 novel *Les Misérables.* In the novel, the protagonist, Jean Valjean, reforms after being released from prison, yet he still spends the rest of his life in hiding for a crime that could be considered justified by the circumstances [29].

Conversely, taken to the extreme, moral relativism sends one slithering down a slippery slope into the quicksand of situational ethics. Niccolò Machiavelli, best known for his treatise on politics, *The Prince,* claimed that situational ethics are justified since they lead to success; in essence, *the end justifies the means.*

> Anyone who determines to act in all circumstances the part of a good man must come to ruin among so many who are not good. Hence, if a prince wishes to maintain himself, he must learn how to be not good, and to use that ability or not as is required. [30]

For Machiavelli, goodness is not intrinsic but relative to results, and sometimes achieving the right goals requires doing behavior that could be considered evil by others. However, in a 1991 conference workshop, John C. Bryan of the University of Cincinnati warned against the dangers lurking behind small, seemingly innocuous ethical compromises made to fit the situation [31].

5.3.3 Contextual Relativism Strikes a Balance Between the Extremes of Blind Absolutism and Situational Ethics

How can individuals avoid these two extremes? Common sense, respect for the rights of humanity, and an internal belief in standards of right and wrong can assist us in deciding what approach to take in resolving an ethical conflict. As David W. Kale observes, "the guiding principle of any universal code of intercultural communication should be to protect the worth and dignity of the human spirit" [32, p. 424]. Ethical behavior entails communicating with people "in a way that does no violence to their concept of themselves or to the dignity and worth of their human spirit" [32, p. 424]. With this focus in mind, he believed that there were some "universal values on which we can build a universal code of ethics in intercultural communication" [32, p. 423]. Such an opinion is echoed by Elvin Hatch, who notes, "... there are certain absolute standards against which all cultures may reasonably be judged, without being guilty of ethnocentrism or intolerance" [33, p. 409]. These humanistic principles, moreover, necessitate that the well-being of people be respected:

> We can judge that human sacrifice, torture, and political repression are wrong, whether they occur in our society or some other. Similarly, it is wrong for a person, whatever society he or she may belong to, to be indifferent toward the suffering of others. People ought to enjoy a reasonable level of material existence: we may judge that poverty, malnutrition, material discomfort, human suffering, and the like are bad. [33, p. 409]

In "Preparing students for the ethical challenges of global citizenship," Dr. Madelyn Flammia proposes a reconciliation of the two viewpoints—*contextual relativism* [26]. According to Ting-Toomey, contextual relativism holds that the "application of ethics can only be understood on a case-by-case and context-by-context basis" [34, p. 4]. In "Values in tension: When is different just different, and when is different wrong?" Thomas Donaldson suggests following these three principles to ethically navigate the treacherous shoals of cultural differences:

- Respect for core human values
- Respect for local traditions
- The belief that context matters when deciding what is right and what is wrong [35]

When an intercultural ethical dilemma occurs in the academic classroom or an industry training session, instructors can start a dialogue to probe the underlying cultural implications of a certain behavior.

After gaining an understanding of the beliefs and values of a different culture, the next step is to seek common ground. Rather than looking at the differences in perspective, the goal is to find common ground [26]. Unless this process unearths a violation of fundamental human rights as discussed earlier, it often leads to a compromise solution within the framework of contextual relativism.

5.3.4 Value Analysis Provides a Systematic Approach to Identifying and Resolving Ethical Conflicts

Value analysis is a systematic process for identifying which values are colliding to create ethical conflicts. This process also allows individuals to define solutions that serve the greater good with the least harm to the stakeholders involved. The process is sufficiently structured to blaze a safe path through a thorny thicket of issues, yet flexible enough to allow for the reality of gray when it comes to such issues.

We believe engineering/science students and professionals are likely to be particularly receptive to value analysis as a tool because their disciplines rest upon the analytical process. In an ethics textbook for undergraduate and graduate students of technical communication, Allen and Voss defined value analysis as a six-step process:

1. Define the *issue* and identify the *stakeholders*
2. Determine the stakeholders' *interests*
3. Identify the relevant *values* that bear on the issue
4. Determine the values and interests that are in *conflict*
5. *Apply a model* to rank values according to importance, to weigh the values and interests that are in conflict
6. *Resolve the conflict* in favor of the higher (more important) value [36, pp. 20–21]

> Value analysis is a systematic process for identifying which values are colliding to create ethical conflicts...The process is sufficiently structured to blaze a safe path through a thorny thicket of issues, yet flexible enough to allow for the reality of gray when it comes to such issues.

Any of the classical models discussed earlier can be used with value analysis, but, as anyone can imagine, the outcome varies considerably depending upon the model used.

To examine these ideas, consider the conflict surrounding the topic of the legalization of drugs. In the Aristotelian model, one would assume that drug use would be considered non-virtuous, a bad habit that would be inimical with achieving happiness and a good life. But applying the utilitarian approach of Mill, one would be led to the conclusion that drug use was an individual decision as long as the use of drugs by one person did not infringe upon the rights of another (such as driving when drunk or stoned). Apply Kant's categorical imperative, on the other hand, and one has to decide on a more-or-less absolute basis whether drug us is "good" or "bad"—given a dichotomous choice, one should think, most likely bad. Introduce the intercultural/international filter and it gets even more complicated (e.g., in Amsterdam, the use of certain drugs in public places is legal, yet the same action in Ankara could lead to a 20-year incarceration).

5.4 Two Value Models Help Rank Values to Resolve Conflicts in Favor of the Greatest Good or the Least Harm

Two common models to define and prioritize ethical values are the concentric-ring and hierarchical models, discussed in this section. Note that in both cases, the values represent a consensus in the US and Western European culture. An extremely important point to note here is that *there is no one formula for emplacing values within this or any other model.* There are consensus values in our culture, and there are consensus values within other cultures (not all the same as ours). There are even basic human values that reflect some consensus across cultures—but no two people, whatever their culture, are going to hold exactly the same values, and in exactly the same order of precedence [36].

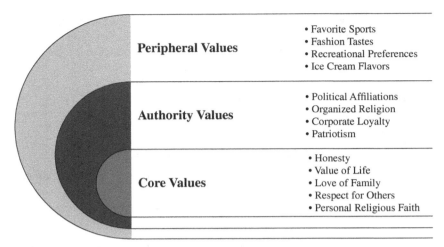

FIGURE 5.2. The concentric-ring model. In the concentric-ring model, the core values are strongest [36].

5.4.1 In the Concentric-Ring Model, Core Values Are Paramount

The personal core values are the strongest, while the peripheral values, reflecting personal preference, are the least strong. In between are the values dictated by various social, political, and religious authorities. A conflict of either core or authority values with peripheral values is usually clear-cut. For example, if a person prefers butter to margarine despite working in the advertising department for Mazola, he or she is not likely to insist the next advertisement feature churnery-fresh butter in lieu of the corporation's product. The concentric-ring model for value analysis is shown in Figure 5.2 [36, p. 24].

Conflicts between core values and authority values get much tougher. Take, for instance, a pediatric surgery that goes bad: the child hemorrhages and needs an emergency blood transfusion to survive, but the parents forbid it on religious grounds. Bound by the Hippocratic Oath, the surgeon orders the transfusion and saves the child's life. The parents sue. In this case, the judge is likely to decide in favor of the surgeon, ruling, in essence, that the child's right to life, a core value, supersedes the parents' religious beliefs, an authority value. If the patient were of age and signed a pre-operation statement refusing a transfusion, the situation becomes much more difficult, because it now becomes a battle of two core values—an adult's personal religious faith versus the surgeon's oath (value of life).

It is quite conceivable for engineers and scientists to encounter similar value conflicts, depending upon which authorities govern their professional lives and which values govern their personal lives. Consider, for example, the conflict an industrial geologist might face if he/she discovered a new petroleum reserve underneath a natural preserve in Montana vital to the ecology of the Missouri River headwaters. In this case, at least two authority values are likely to collide: as an employee of a major oil company, the geologist is clearly obligated to report the find in the company's business interest. But is the geologist also ethically obligated as a loyal citizen to share the discovery with the responsible governing agencies so they can take action to block the company's attempt to convert the new find to profit? It gets even more complicated to consider the politico-economic imperative for the nation to achieve energy independence. Other authority values could enter the picture, such as the geologist's political affiliations, her/his attitude toward environmental protection groups, etc. There are often no easy answers to conflicts of this nature. That's why it's useful to have more than one value model to work with.

5.4.2 The Hierarchical Model Highlights Intense Conflicts Between Legal, Societal, and Authority-Based Values

In this case, the strongest values, critical shared values, are on top, followed by important shared values, conscientiously shared values, and personal preference values. The latter resemble the peripheral values in the outer ring of the concentric-ring model, and the conscientiously shared values are much like the authority values in the middle ring. The hierarchical model for value analysis is shown in Figure 5.3 [36, p. 27].

The core values in the concentric-ring model are divided into the top two tiers in the hierarchical model. Both critical and important shared values are consensus values in Western culture; the difference is that the former are backed by law whereas the latter are "enforced" via societal pressure. Thus, if a person commits murder and is caught, he

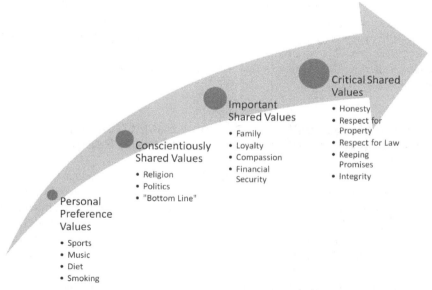

FIGURE 5.3. The hierarchical value model. In the hierarchical value model, the critical shared values are backed by law, whereas the important shared values are "enforced" only by social opinion [36].

or she is likely to incur severe penalties under law, whereas if a person refuses a helping hand to a friend or family member in need, he or she is likely to be much less liked, possibly even ostracized, but not likely to be arrested.

As with the concentric-ring model, conflicts between personal preference values and any of the three higher levels are generally easy to resolve. For example, to tie in with Mill's utilitarianism, smokers' right to smoke (personal preference) ends where their neighbors' lungs (physical integrity) begin.

Conflicts between critical and important shared values can be very difficult. Consider, for example, how serious a crime a close family member would have to commit before someone would turn him or her into the police. How many spouses have performed a citizen's arrest on their partner for a trivial offense like going five miles over the speed limit? At the opposite extreme, most family members would feel morally and legally obliged to turn in a sibling who was an ax murderer. But what if a person's brother were embezzling, say, $5,000 a year from his filthy-rich boss (who underpays and mistreats him) to provide medical care for their niece, who has a disability? At the extremes, it's clear. In the middle, it gets tough.

Similarly, engineers and scientists may have to walk a tightrope between critical shared values, important shared values, and conscientiously shared values, particularly if they work for large corporations whose conscientiously shared value is the "bottom line." That works unless the company takes liberties with the law in the interest of profit. Respect for the law, a critical shared value, demands an engineer or scientist who is aware of such a situation report it to the cognizant government authority—to "blow the whistle," so to speak. If the government succeeds in enforcing the law, the engineer or scientist is legally protected against retribution from her/his employer. But what if the corporate lawyers get the company off the hook? How much longer will the engineer or scientist be receiving a paycheck? Or, for that matter, finding another job within the industry? Fear of such a possible outcome gives many potential whistle-blowers pause, because another important shared value—financial security for self and family—enters the picture.

5.5 Value Models within Technology-Based Companies and Professional Associations Offer Broad Ethical Perspectives

It makes sense that the ethical codes and guidelines of professional organizations within the technical community and those of high-technology companies would provide a useful framework for value analysis specifically oriented to the ethics of intercultural communication within an engineering or scientific context. Figures 5.4, 5.5, 5.6, and 5.7 present four such value models for the Society for Technical Communication® (STC), the Institute of Electrical and Electronics Engineers® (IEEE), Lockheed Martin Corporation®, and Harris Corporation®, respectively [37, 38, 39, and 40]. (For website links to the full text on these ethical guidelines and codes, see the References list.)

To better understand how these four codes, as well as the concentric-ring and hierarchical value models, can help shed light on ethical conflicts within intercultural communication, consider the following example. A civil engineering company is hired to design a bridge in a developing nation. The government of that nation orders the

VALUE MODELS WITHIN TECHNOLOGY-BASED COMPANIES

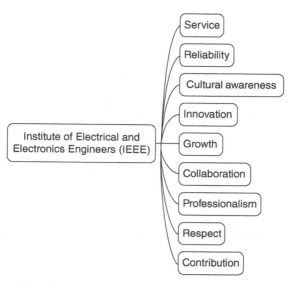

FIGURE 5.4. The IEEE Code of Ethics. The IEEE Code of Ethics focuses on the integrity of the profession, which includes cultural awareness (authors' interpretation) [37].

company to reduce the strength specifications for the steel girders in order to award the girder contract to a local company as part of an industrial offset agreement within the contract for the bridge.

Applying the STC ethical guidelines, both quality and legality (US law) dictate the strength specs not be altered, because doing so could pose a threat to human life due to a bridge collapse during an earthquake. Applying the IEEE core values, the values of

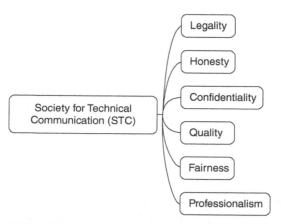

FIGURE 5.5. The STC's Ethical Principles. Within the STC's Ethical Principles, cultural sensitivity falls under fairness (authors' interpretation) [38].

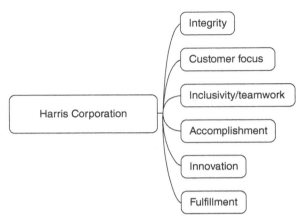

FIGURE 5.6. The Harris Corporation Code of Conduct. In the Harris Corporation Code of Conduct, inclusivity and customer focus both demand cultural sensitivity (authors' interpretation) [39].

service and reliability demand the company stand firm on the original specs, both for the public good in the host country and for the reputation of the engineering profession. Although neither Lockheed Martin nor Harris builds bridges, applying their corporate codes of ethics offers an interesting perspective on the situation. Both companies' codes call for customer focus, which, strictly construed, could point toward reducing the specs for the girders to be responsive to a request from the government of the host country so as to comply with the terms of the contract. A bridge-building company would also be likely

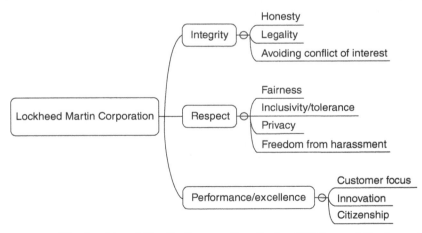

FIGURE 5.7. The Lockheed Martin Corporation Code of Ethics and Business Conduct. In the Lockheed Martin Corporation Code of Ethics and Business Conduct, performance is measured in customer satisfaction, and inclusivity and tolerance fall under respect (authors' interpretation) [40].

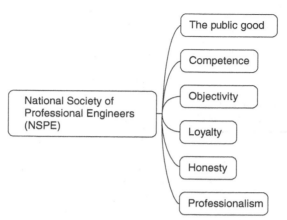

FIGURE 5.8. The NSPE Code of Ethics. The NSPE Code of Ethics places the greatest emphasis on the public good without specifically addressing intercultural conflicts (authors' interpretation) [41].

to have customer satisfaction as a core ethical tenet. However, for Lockheed Martin and Harris, other core values (accomplishment for Harris, citizenship for Lockheed Martin, and integrity for both) would overrule the other considerations. It is also likely that in this context, "customer focus" would be applied to the *end user* (i.e., the citizens driving across the bridge who would be killed if the girders failed during an earthquake) rather than to the contractual customer, the host government.

Further insight into this scenario can be gained by applying the ethical codes of the National Society of Professional Engineers® (NSPE) and the American Society of Civil Engineers® (ASCE), which run along similar lines (refer to Figures 5.8 and 5.9) [41], [42]. Both codes assign top priority to defending the public good; clearly, these two professional engineering associations would call for the engineers designing the bridge to refuse to jeopardize public safety by compromising the structural integrity of the bridge.

And although neither the NSPE nor the ASCE code directly addresses the intercultural elements of ethical values, the latter—from the organization most directly relevant to bridge construction—drives to the heart of the matter by explicitly calling for "zero tolerance for bribery, fraud, and corruption" [42]. That, in turn, leads to a fairly common cause of ethical conflicts in doing business on the international market—what is considered illegal in one country or culture may be common practice in another.

5.6 Before Analyzing Ethical Conflicts in an Intercultural Context, It's Important to Understand the Cultural Differences Involved

As evident in the previous discussion, although value analysis does impose structure on an inherently subjective process of weighing values and interests to arrive at "do the most good, do the least harm" solutions in situations of ethical conflict, things can still

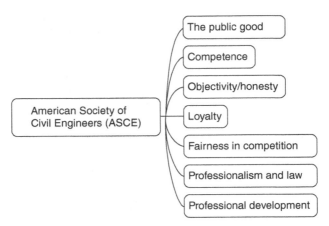

FIGURE 5.9. The ASCE Code of Ethics. The ASCE Code of Ethics, like that of the NSPE, charges engineers to pursue actions that are for the greater good of the public (authors' interpretation) [42].

get complicated. That is because the fault lines separating the tectonic plates of critical and important shared values (not to mention the highly unstable intersection of fiercely defended authority-based values) are seismic time bombs—frequently generating "tremblers" and sometimes even all-up earthquakes. Such conflicts are difficult enough within a homogenous culture, but in the melting-pot culture of the United States, they can lead to major earthquakes.

Toss in the intercultural value filter, and the picture gets even more complex. That's because while ethical constructs such as the concentric-ring and hierarchical model discussed earlier point toward a general consensus on critical and important shared values in the United States and Western Europe, there can be substantial differences when introducing value systems from Eastern Europe, Southeast Asia, the Middle East, sub-Saharan Africa, and elsewhere. And when intractable authority-based values such as religious affiliations come into conflict across nations and cultures, tensions can become high.

The authors are not so arrogant as to claim subject matter expertise in all these diverse cultures. Indeed, that raises an important point in teaching the ethics of intercultural communication to a technical audience—instructors cannot meaningfully introduce the intercultural value filter without themselves understanding the concepts of intercultural communication. Fortunately, the collective wisdom of all the authors who contributed to this anthology vastly exceeds ours, so we can focus on the *process* of analyzing ethical conflicts across the values of different cultures rather than be concerned about our modest degree of knowledgeability of the many different cultures.

Similarly, neither engineering professors nor industry trainers need to be walking encyclopedias on the cultural variations of hundreds of nations to teach the basics of ethics and intercultural communication. However, in approaching specific instances of cross-cultural ethical conflicts—be that in a classroom exercise or in a real-world business situation—it's critical to stress the importance of learning about the cultures in question before attempting to use value analysis to identify and resolve the conflicts. To do otherwise is to fall into the trap of ethnocentrism and stereotyping—distorting the values of other cultures by viewing them through the prism of our own. Open-mindedness and cultural sensitivity are essential characteristics in using value analysis to objectively identify and resolve ethical conflicts across international and intercultural lines.

5.6.1 Conflicts Between Authority-Based Values (Corporate, Political, National, and Religious) Can Be Explosive

To put some "form and shape" to this rather amorphous concept, let's revisit the six value models we just described, adding an intercultural filter. In both the concentric-ring and hierarchical value models, honesty is generally considered the most central core value or the highest critical shared value. This is probably also the closest we will get to a "universal" value; whatever differences they may have in other areas, most cultural value systems consider honesty and integrity to be of paramount importance. That's why dishonesty and insincerity can somehow usually be detected even across linguistic and cultural barriers, both verbal and non-verbal. Most human beings have an innate sense of what's right and wrong, what's fair and unfair, and what's honest and dishonest, that crosses cultural lines [36].

It is at the opposite end of the spectrum (peripheral or personal preference values) that we are apt to find the most immediately noticeable differences across cultures (e.g., football versus soccer, pasta versus couscous, opera versus reggae). Fortunately, while such personal preferences may be strongly felt, for the most part, there is an inborn intercultural tolerance—even curiosity and willingness to experiment across cultural lines. Thus, World War III is unlikely to result from a dispute over whether the Super Bowl or the World Cup is the most important athletic title.

But at the higher (or more central) value areas, cross-cultural differences can become much more serious. We have posited that, for the most part, "honesty is honesty," across cultural lines. But try law. While there is such a thing as international law, what agent is going to enforce it, particularly if the violator has more political/military clout than the enforcing agency, such as the World Court or the United Nations? In an increasingly global economic marketplace, engineers and scientists and the companies they work for are more likely to have to deal with significant variations in law as they work in different areas of the world. Indeed, the most successful international corporations train their technical, management, and marketing personnel in the nuances of intercultural law, customs, and value systems because they have come to realize that cultural awareness and sensitivity walk hand in hand with business success.

A US company faced an ironic situation involving child labor at a plant it acquired in Southeast Asia. In compliance with US law, they put an end to the practice (which was perfectly legal in the nation in question). Imagine the company's consternation when they learned that dismissing the children had created a financial hardship for many of their families, as their income was a key part of the household budget. In an enlightened intercultural move, the US company rehired the children but they worked only half a day, the other half being devoted to their schooling [13].

Love of family crosses cultures but the dynamics within families differ radically. For example, in business, professionals may witness the dominant role of men in many

Arabic nations versus the matriarchal societies in certain parts of Africa or even within some Native American cultures. A professional colleague tells how female friend (we'll call her Terry) faced an international business situation where her leadership position was challenged based on gender and how Terry's company (we'll call it "Bild-It") handled the situation.

> Bild-It had just won a bid to work with a South American engineering firm (we'll call it BigDeal, Inc.) on a major construction project—one of the largest projects to date for both companies. There was much buzz about this new project, and Bild-It's reputation was riding on smooth collaboration with BigDeal and the success of the overall project. BigDeal had much to win or lose with this contract as well.
>
> Bild-It chose Terry to be the project lead. She had seniority, she had the experience, and she specialized in large construction management. As a side note, her name was one that could have been for a female or a male. As it turns out, BigDeal assumed Terry was a male.
>
> The Bild-It team flew to company headquarters in South America. Upon Terry's first face-to-face meeting with the team from BigDeal, the project manager pulled the Bild-It CEO (a male) aside and informed him that BigDeal was an engineering firm "with a good reputation"; therefore, their engineers would not work with women. BigDeal threatened to pull from the partnership if Terry remained the lead.
>
> And, in an amazing move that surprised even Terry, the CEO said, in essence, "Well, then we will have to leave." BigDeal couldn't believe it, and immediately back-pedaled. The female engineer stayed on, and she did great work. Only later did Bild-It find out it wasn't BigDeal's company policy that opposed a female lead, nor was it culture. It was just one male employee who didn't want to work with females. The US company had assumed for a long time it was a cultural issue they encountered that day, but it wasn't at all [43].

The lesson learned here is that in dealing with ethical conflicts across cultures, it is wise never to make assumptions based on expectations. Each case is different. Many do run along expected lines, but some take an unexpected 180-degree turn. Before a conflict can be resolved, it's important to identify all the stakeholders and understand the values and interests that motivate them.

The most intense situations of cross-cultural conflict generally result not *across* hierarchical value levels (or across "rings"), rather *within* the third level in the hierarchy, conscientiously shared values or inside the corresponding middle ring, authority values. This is true even within the US/Western culture, and the differences become even more marked across cultures. Consider four typical authorities (and corresponding allegiances people are likely to have):

- Corporate affiliation (such as within the fiercely competitive aerospace industry where many engineers find employment)

- Political parties
- National loyalties
- Organized religions

One could make a case that the intensity of internal conflicts within those four authority areas increases in the same order they are presented. The opportunity for conflict between authority values in transacting business across "party lines," across intercultural and international lines, and across religious belief systems is extensive. The list is endless: Democrat versus Republican, Whig versus Tory, Greek versus Turk, South Korean versus North Korean, Catholic versus Protestant, Jew versus Muslim. It has been said that in the recorded history of humankind, more deaths have resulted from religious conflict (or national conflicts rooted in religious differences) and religious persecution than from any other cause [44, 45]. As Mark Twain once said, "[Man] has made a graveyard of the globe in trying his honest best to smooth his brother's path to happiness and heaven" [46, p. 121].

5.6.2 Examining Models of Ethics in Cross-Cultural Contexts Often Pivots on the Precarious Balance of Relativism and Universalism

The intercultural filter can be applied to any value model being used in value analysis. For example, consider the ethical codes of the four technical communities discussed earlier. Applying the STC values in an intercultural context, one can readily see where legality would have different implications in different countries, and confidentiality/privacy would take on different implications in a high-context Eastern culture than in a low-context Western one. For example, a computer program that designed a nasty buzzing sound in the case of operator error was very popular in the United States, but as Hoft reported, sales were abysmal in Japan [13]. Ebullient "Yankees" don't mind advertising our screw-ups, but for the Japanese user, calling attention to a mistake would be "losing face."

Turning to the IEEE ethical values, while they do include cultural awareness, that very awareness would make it evident that other core values such as collaboration and respect might play differently in Sri Lanka than in Omaha. Lockheed Martin's values include a specific directive to "honor the laws of the nations in which we do business," but how does that work in a Third World nation where what US business—and lawmakers—would call "kickbacks" to win contracts are not only legal but an expected part of the business process [39]? We asked this question to James Messina, ethics director at Lockheed Martin Missiles and Fire Control in Orlando, and he said that in such a case, US law and the corporation's core value of integrity would take precedence [47]. Applying the Harris Corporation ethical code to that situation would lead to a similar conflict between inclusivity/teamwork and integrity.

In essence, the concept "when in Rome do as the Romans do," which suggests cultural/moral relativism, collides with fundamental values such as integrity, which suggests cultural/moral universalism. In fact, the NSPE code of ethics included a "When in Rome" clause in 1966 stipulating that it was permissible to provide "tenders for work

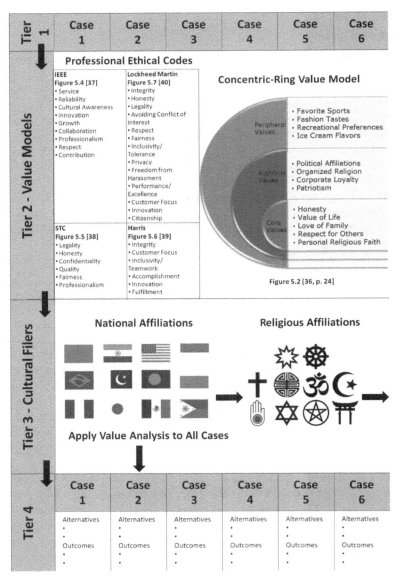

FIGURE 5.10. Process flow for value analysis. A systematic process flow for value analysis applies the perspective of multiple value models as well as cultural differences to identify and resolve ethical conflicts in intercultural technical communication [36].

in foreign countries when such is required by law, regulations, or practices of the foreign country" [41]. However, realizing the potential problems this relativist approach to ethics could cause and deciding to occupy a more "pure position on competitive bidding," the NSPE rescinded the clause two years later [48]. This perfectly exemplifies the difficulty that ethical standards have in resolving the tension between cultural/moral

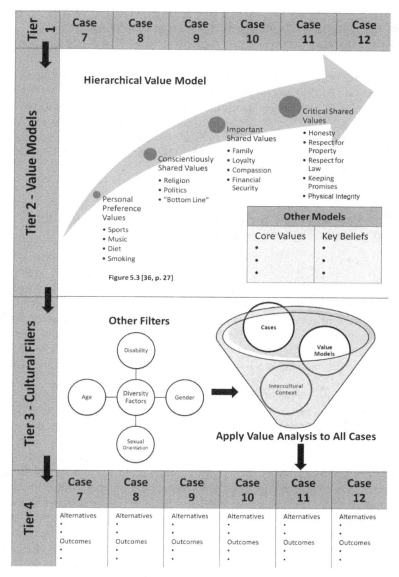

FIGURE 5.10. Continued

relativism and cultural/moral universalism. The discussion of the ideas in the next section (Section 5.7) further examines this philosophical dichotomy, because it is at the epicenter of many of the ethical earthquakes that occur along the fault lines between different cultures. As we depict and describe the process flow for value analysis in Section 5.7, we apply all the filters we have been describing to identify and resolve ethical conflicts between different value models, including different cultures.

5.7 Analyzing Case Histories via a Multi-Tiered Process of Ethical Models and Cultural Filters that Clarifies Ethical Conflicts, Defines Alternative Actions, and Predicts Outcomes

Engineers and scientists, being analytical creatures by nature, may find the process flow depicted in Figure 5.10 on pages 114–115 to be an intriguing approach to identifying and resolving ethical conflicts in intercultural communication within the technical field.

The figure, adapted from a similar illustration developed by Allen and Voss [36, p. 342], is designed to examine different international ethical situations individuals might encounter.[1] Methods for applying this model will be examined in the following components of this section.

5.7.1 The Process Flow for Value Analysis is a Complex but Systematic Chart: Perfect for Engineers

Starting with an ethical conflict scenario such as Cases 1-12 in Tier 1, the idea is to apply the six value models shown in Tier 2 of Figure 5.10 to the situation in question to identify the values and interests of the stakeholders and determine which ones are in conflict. When applying the "filters" of the hierarchical and concentric-ring models, the point is to *search for potential actions that resolve the conflict in favor of the higher or more central value.* Similarly, viewing the situation through the ethical lens of each of the four organizations to discern other areas of ethical conflict and explore options can help resolve the conflicts in a manner that serves the greatest good and/or does the least harm.

The seventh value model at the right-hand side of Tier 2 is deliberately left empty. This is because it is a part of the exercise where the trainer can insert or add another, different (i.e., "wild card") ethical code or value model relevant to that discipline, course, or industry training program. For example, the earlier discussion of the "faulty bridge" scenario provided further insight into the value conflict by applying the ethical codes of the NSPE and ASCE. Moreover, no one is required to use any of the six models in the figure; rather, the same process will work with any other ethical models. This factor is important, for it is generally useful to apply multiple models because different perspectives work better with different situations. Accordingly, the trainer can introduce personal values in the "wild card" model. In so doing, however, it is important that the trainer apply such wild card models *independently* of the other ethical codes and value models, for doing so prevents them from influencing the interpretations that result from applying the other models to the situation being examined.

Tier 3 of Figure 5.10 is the central or the core idea of the topic—the intercultural "filter" through which the analytical conclusions reached in the second tier must be processed to account for cross-cultural differences in values and priorities. On the left side

---

[1] For a printable 11×17 pdf of Figure 5.10 and details on the case studies, contact the authors Dan Voss (danvoss999@gmail.com) or Bethany Aguad (bethanyaguad@gmail.com).

of Tier 3, the flag icons represent the diversity of international cultures. Closely associated with those elements are religious belief systems, depicted iconically left of center. And if the professor or trainer wishes to apply a broader definition of "culture," he/she can introduce other factors like those shown to the right of center (gender, age, sexual orientation, disabilities). Such an approach would make this process model extremely useful in company in-service education programs on diversity, tolerance, harassment, and equal opportunity.

At the right-hand side of Tier 3 of Figure 5.10, the "funnel" represents the value analysis process as the various ethical codes, value models, and cultural filters are applied to the ethical conflict under review. Methods for applying this model will be examined in the following components of this section. Included within the funnel is the personal value system of each student (and, for that matter, the professor or trainer). It's important to remind students to do their best to "suspend" their personal value systems when applying the depicted ethical codes, value models, and cultural filters, but it is human nature for one's personal values to influence the process. Rather than have students' personal values muddy the waters when applying various value models and intercultural filters to the situation, urge them to apply those *after* completing the earlier steps in the process, as a "sanity test" on the proposed alternative actions to resolve the ethical conflict in in question.

The group exercise concludes with Tier 4 of Figure 5.10: defining a number of possible alternatives for each case history and—this is important—projecting the likely outcome of each action. The lesson learned there is like a law of physics: just as every action has an equal and opposite reaction, every decision has consequences. Sometimes choosing the highest or most central ethical values can take a person down a path that has severe negative consequences. Conversely, sometimes taking a less well-defined and problematic ethical path involving situational ethics (i.e., making the rules up to serve one's personal interests) can be very tempting, because it leads to positive personal outcomes rather than being penalized for doing what is right.

5.7.2 It's Tempting to Quantify the Ethics of Intercultural Communication into an Equation. All Suggestions Welcome!

When Allen and Voss conceived this diagram and applied it to twelve case histories involving ethical conflicts within technical communication, they toyed with the idea of trying to *quantify* the process. To do so, they

- First assigned numerical values to ethical values
- Arithmetically weighted these numerical values to allow for the interests of the various stakeholders (as well as their "power" to influence the outcome)
- Finally developed a formula for computing the "ethics quotient" that results by applying different models to the conflicts in a case

Through this approach, Allen and Voss soon discovered the possibilities were endless. In fact, for each formula they explored, the outcome always resulted in the numerical values assigned to the ethical values and the percentages of arithmetic weighting—both of which were arbitrary and therefore subjective. From these findings, Allen and Voss concluded that while the process could be fascinating, the search for a quantitative analytical outcome from an intrinsically qualitative process was futile.

5.8 Suggestions for Integrating the Specialized Topic of this Chapter into Academic Courses and Industry Training Classes

In this section, we offer suggestions, resources, and materials that may be helpful to college professors and industry trainers who are teaching intercultural communication to a technical audience, including students majoring in science or engineering as well as practicing scientists and engineers. We provide methodology and resources appropriate to three teaching/training situations:

- A professor who is a subject matter expert (SME) in intercultural communication teaching a class on that subject to engineering or science students
- A professor who is an SME in science or engineering including an instructional module on intercultural communication within an engineering or science course
- An industry trainer who is an SME in intercultural communication teaching an in-service education class, workshop, seminar, or webinar for engineers or scientists

5.8.1 Situation 1: SME Professor in Intercultural Communication Teaching Engineering or Science Students

Given the narrowness of our topic in this chapter, it is unlikely that a full-semester college course focusing exclusively on the ethics of intercultural communication would be part of the curriculum in either the sciences or the humanities. Assuming a typical one-semester course on intercultural communication consisting of sixteen 1-hour classes, the ethics "module" would probably consist of just one or two sessions. Such a module would fit best near the end of the course so that students will already be familiar with the basics of intercultural communication and therefore ready to view the subject through the "zoom lens" of ethics and cross-cultural value systems (see Figure 5.11).

With no more than two 1-hour class sessions likely to be devoted to the ethics of intercultural communication, the best treatment would probably be to assign reading on ethical models and value analysis for the first class, which would consist of a 30- to 40-minute lecture on the subject supported by a PowerPoint® presentation and handouts, followed by questions-and-answers and discussion.

The second class would consist of a "hands-on" exercise in value analysis. Given the complexity of the subject and the need to bring different value models to bear on cross-cultural ethical conflicts, it would generally be best to have the students work in teams, with each team analyzing one case and presenting a 2- to 3-minute summary to the class. To fit this exercise in a 1-hour window, it would be a good idea to assign

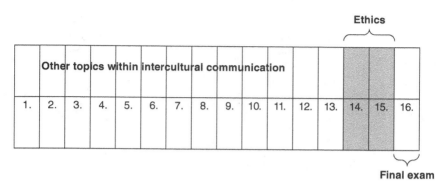

FIGURE 5.11. Situation 1: Ethics unit in an intercultural communication course. The ethics of intercultural technical communication would likely occupy just one or two sessions in a one-semester course in intercultural communication for engineers/scientists.

the students to read the cases *before* class and have each choose one case and submit a brief analysis at the beginning of class as a graded assignment. That would ensure the teams are spun up and ready to collectively work one or two case analyses within 30 to 35 minutes, allowing time for each to give a brief report to the class. Depending upon the nature of the final exam, it would be appropriate to include a few questions on this topic.

5.8.2 Situation 2: SME Professor in Engineering or Science Teaching a Technical Content Course

The other likely educational scenario would be for a professor of engineering or science to include a module on intercultural communication within a content course in engineering or science. Here the window of class time to be devoted to a narrow subset of that topic—the ethics of intercultural communication—would be much smaller, since the module on intercultural communication itself would be unlikely to occupy more than one or two 1-hour class sessions (Figure 5.12), most likely near the end of the course after the technical content has been covered.

Realistically, that would leave just a 30-minute segment of class time for the ethics of intercultural communication. To explain and illustrate the concept of using value analysis to resolve cross-cultural ethical conflicts would most likely require the full 30 minutes. Even with advance reading, attempting to cover this topic in less time than that would limit the coverage to a lecture/presentation, with little or no time for discussion. If managed efficiently, a 30-minute segment would allow for a 15-minute overview, after which the instructor could lead the class through value analysis of two or three cases involving cross-cultural interactions and ethics. If the students were then assigned a separate case to analyze as a follow-up assignment (or possibly as a final exam question), that would allow meaningful coverage of the topic in the limited time available.

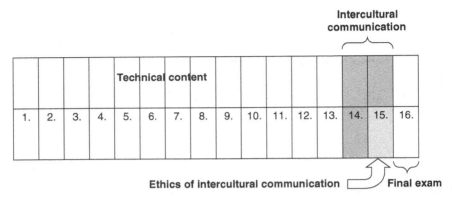

FIGURE 5.12. Situation 2: Ethics module in intercultural communication unit in an engineering or science course. The ethics of intercultural technical communication would likely be a 30-minute module within a two-lesson unit on intercultural communication in a one-semester engineering or scientific content course.

5.8.3 Situation 3: Industry In-Service Education Class or Training Workshop in Intercultural Communication

Most large high-tech companies have formal employee training programs in "soft" subjects such as ethics, diversity, and intercultural communication as well as in the technical disciplines. A variety of formats are used for industry training programs.

- Classroom instruction in a mini-course format, typically just one to four 1-hour sessions
- Computer-based training in modules of approximately one hour
- Seminars or workshops ranging from a couple hours to a full day
- Webinars, typically one to two hours.

Whatever training format or combination of formats a company uses, just as in academe it is unlikely a company would devote an entire training course exclusively to the ethics of intercultural communication. However, the natural overlap of ethics, diversity, and intercultural communication creates opportunities to focus on this topic (see Figure 5.13).

Of these three curriculum models, the most thorough treatment of the ethics of intercultural communication would probably be within an ethics course/workshop (Figure 5.13a), since cultural sensitivity, tolerance, diversity, and equal opportunity are all values commonly found in companies' ethical codes or guidelines. Indeed, it's hard to imagine an industry ethics course that did *not* venture into these areas.

It would also be possible to include an ethics module within a course on diversity or intercultural communication, but in those models (Figures 5.13b and 5.13c) the treatment of the sub-topic of ethics in intercultural communication would probably be more

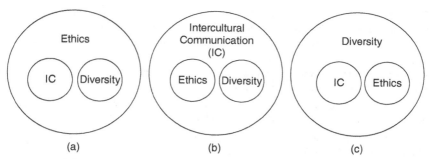

FIGURE 5.13. Situation 3: Integrating intercultural ethics into industry training courses. In industry training, the most thorough treatment of the ethics of intercultural technical communication would likely occur in the "intercultural communication" and "diversity" modules of an ethics course (a), with "lighter" treatment in the ethics module of a course on intercultural communication (b), or diversity (c).

limited, since ethics would typically constitute a smaller percentage of the curriculum in a diversity or intercultural communication class than diversity or intercultural communication would occupy within the curriculum of an ethics course. In the former case (Figure 5.13a), it should be possible to provide both instruction and a group exercise in the ethics of intercultural communication, whereas in the latter (Figures 5.13b and 5.13c) the treatment of ethics in intercultural communication would tend to be more limited.

5.9 Conclusion: The Authors Invite Further Research and Contributions

In conclusion, teaching the ethics of intercultural communication to a technical audience, whether it is composed of students or professionals, is a rather specialized educational niche. A review of the literature indicates that although abundant instructional materials are available for teaching ethics and intercultural communication separately, very little research and resources exist for teaching the two together. It is our hope that the discussion presented in this chapter will spark further interest in this specialized but crucial area of intercultural communication.

References

1. H. Zandvoort, "Preparing engineers for social responsibility," *European Journal of Engineering Education*, vol. 33, no. 4, pp. 133–140, 2008.
2. M. J. Riemer, "Intercultural communication considerations in engineering education," *Global Journal of Engineering Education*, vol. 11, no. 2, pp. 197–206, 2007.
3. P. M. Dombrowski, private communication, February 2012.
4. H. Luegenbiehl, "Ethical autonomy and eng. in a cross-cultural context," *Techné: Research in Philosophy and Technology*, vol. 8, no. 1, pp. 4–10, Fall 2004.

5. D. Radebaugh, K. Soschinske, G. Rimmington, and M. Alagic, "A web-based interactive approach for engineering ethics training in a global learning environment," in *Proceedings of Midwest Section Conference of the American Association for Engineering Education*, Kansas City, 2006, p. 4.
6. B. Newberry, "The dilemma of ethics in engineering education," *Science and Engineering Ethics*, vol. 10, no. 2, pp. 343–351, 2004.
7. C. E. Harris, Jr., "Explaining disasters: The case for preventative ethics," *IEEE Technology and Society Magazine*, vol. 14, no. 2, pp. 22–27, 1995.
8. C. E. Harris, Jr., "The good engineer: Giving virtue its due in engineering ethics," *Science and Engineering Ethics*, pp. 153–164, 2008.
9. R. V. Ramani, "Engineering ethics: An area in need of greater understanding," *Mining Engineering*, vol. 63, no. 8, pp. 55–67, 2011.
10. P. M. Dombrowski, "Can ethics be technologized? Lessons from *Challenger*, philosophy, and rhetoric," *IEEE Transactions on Professional Communication*, vol. 38, no. 3, pp. 146–151, September 1995.
11. G. Geistauts, E. Baker, and T. Eschenbach, "Engineering ethics: A system dynamic approach," *Engineering Management Journal*, vol. 20, no. 3, pp. 21–28, 2008.
12. G. Hashemian and M. C. Loui, "Can instruction in engineering ethics change students' feelings about professional responsibility?" *Science and Engineering Ethics*, pp. 201–215, 2009.
13. M. Flammia and D. Voss, "Ethical and intercultural challenges for technical communicators and managers in a shrinking global marketplace," *Technical Communication*, vol. 54, no. 1, pp. 72–87, 2007.
14. P. Zemliansky, private communication, February 2012.
15. E. de Graaf, and W. Ravesteijn, "Engineering. Education.: From competencies to training methods," *The Many Facets of International Education of Engineers*, pp. 1–6, 2000.
16. N. Escudeiro, P. Escudeiro, A. Borata, and C. Lobo, "Enhancing students' teamwork and communication skills in international settings," in *Proceedings of Information Technology Based Higher Education and Training 2011 International Conference*, Kusadasi, Izmir, 2011, pp. 1-8.
17. P. M. Dombrowski, *Ethics in Technical Communication*, Boston, MA: Allyn and Bacon, 2000, p. 5.
18. C. Susskind, "An engineer's Hippocratic Oath," *Understanding Technology*, 1973, p. 118.
19. Aristotle, *Nicomachean Ethics*. Available at: http://nothingistic.org/library/aristotle/nicomachean/nicomachean01.html (accessed February 15, 2015).
20. Aristotle, *Nicomachean Ethics*. Available at: http://nothingistic.org/library/aristotle/nicomachean/ (accessed February 15, 2015).
21. Confucius. (1996). *Lun Yu: The wisdom of Confucius* [Translation by N. P. Tsui]. Available at: http://www.confucius.org/lunyu/lange.htm (accessed November 1, 2013).
22. M. Sim, "The moral self in Confucius and Aristotle," in *Remastering Morals with Aristotle and Confucius*, Cambridge, MA: Cambridge University Press, 2007, pp. 134–165.
23. J. Bentham, *A Fragment on Government*, London, 1776, para. 2. Available at: http://www.efm.bris.ac.uk/het/bentham/government.htm (accessed January 1, 2016).
24. J. S. Mill, *Utilitarianism*, London: Oxford University Press, 1998, p. 55.
25. I. Kant, *The Metaphysics of Morals* [Translation by M. J. Gregor], Cambridge, MA: Cambridge University Press, 1996.

REFERENCES

26. M. Flammia, "Preparing students for the ethical challenges of global citizenship," in *Proceedings of 9th International Conference on Politics and Information Systems, Technologies and Applications: PISTA 2011.* 2011 © IIIS. Available at: http://www.iiis.org/CDs2011/CD2011SCI/PISTA_2011/Abstract.asp?myurl=PA799AM.pdf (accessed November 4, 2013).
27. C. Fink, *Media Ethics*, Boston, MA: Allyn and Bacon, 1988, p. 7.
28. F. Boas, "Museums of ethnology and their classification," *Science*, 1887, p. 589.
29. Encyclopædia Britannica Inc., *Les Misérables*. Available at: http://www.britannica.com/EBchecked/topic/385213/Les-Miserables (accessed November 4, 2013).
30. Political Theory from Classical Times to the Renaissance. Available at: http://www.stjohnschs.org/english/renaissance/ren-pol.html (accessed February 15, 2015).
31. D. Voss, "A slippery slope: Using value analysis to resolve ethical conflicts in technical communication," *Proceedings of 40th Annual International Conference of the Society for Technical Communication,* Dallas, 1993, pp. 288–291.
32. D. W. Kale, "Peace as an ethic for intercultural communication," in *Intercultural Communication, A Reader*, Belmont, CA: Wadsworth Publishing, 1994, pp. 437–438.
33. E. Hatch, "The evaluation of culture," in *Intercultural Communication, A Reader*, Belmont, CA: Wadsworth Publishing, 1994, p. 409.
34. S. Ting-Toomey, *Communicating Across Cultures*, New York: The Guildford Press, 1999, p. 4.
35. T. Donaldson, "Values in tension: When is different just different, and when is different wrong?" *Harvard Business Review*, pp. 48–61, 1996.
36. L. Allen and D. Voss, *Ethics in Technical Communication: Shades of Gray*. New York: John Wiley & Sons, Inc., 1997.
37. Institute of Electrical and Electronics Engineers, 2013. *7.8 IEEE Code of Ethics*. Available at: http://www.ieee.org/about/corporate/governance/p7-8.html (accessed November 4, 2013).
38. Society for Technical Communication, March 20, 2007. *Introducing STC Ethical Guidelines*. Available at: http://www.stcsig.org/cic/onlinebook/chapter_22/c22a.htm (accessed November 4, 2013).
39. Lockheed Martin, 2007. *The Ethics Program.* Available at: http://www.lockheedmartin.com/us/who-we-are/ethics/ethics-program.html (accessed November 4, 2013).
40. Harris Corporation, 2012. *Harris Mission and Values.* Available at: http://www.harris.com/mission-values.asp (accessed November 4, 2013).
41. National Society of Professional Engineers, 2007. *NSPE Code of Ethics for Engineers.* Available at: http://www.nspe.org/resources/pdfs/Ethics/CodeofEthics/Code-2007-July.pdf (accessed November 4, 2013).
42. American Society of Civil Engineers, 2013. *Code of Ethics.* Available at: http://www.asce.org/Ethics/Code-of-Ethics/ (accessed November 4, 2013).
43. T. Nathans-Kelly, private communication, October 2013.
44. S. Kingsmore (2009, October 17). *Christianity and Religion Have Caused More Deaths Than Anything Else in History*. Available at: http://scottfromsc.blogspot.com/2009/10/christianity-and-religion-have-caused.html (accessed November 4, 2013).
45. ProvetheBible.net, 2010. *Isn't Religion to Blame for Most of History's Killings?* Available at: http://www.provethebible.net/T2-Objec/G-0101.htm (accessed November 4, 2013).

46. M. Twain, *Mark Twain's Book of Animals*. Berkley: University of California Press, 2009, p. 121.
47. J. Messina, private communication, October 2013.
48. National Society of Professional Engineers, 1976. *Report on a Case by the Board of Ethical Review*. Available at: http://www.nspe.org/resources/pdfs/Ethics/EthicsResources/EthicsCaseSearch/1976/BER%2076-6.pdf (accessed November 4, 2013).

Online Contexts

6

Autonomous Learning and New Possibilities for Intercultural Communication in Online Higher Education in Mexico

César Correa Arias

University of Guadalajara

Education is a heterogeneous worldwide system that is in constant transformation due to continuous social, economical and technical changes and challenges. Such factors continually confront contemporary societies when they try to figure out what could be the role of technologies in transforming educational processes. For nearly three decades, information and communication technologies (ICTs) have been changing the way teachers and scholars experience didactics and curriculum processes inside institutions of higher education. However, at the moment of integrating ICTs in the curriculum, as an institutional policy, there could be significant fissures between curricular and didactic purposes, applied methodologies and educational practices. It is the coherence between curriculum and didactic development that could give ICTs, as mediation, an important pedagogical role within the processes of learning and teaching. This chapter examines the role of autonomous learning among students and teachers mediated by ICTs in building meaningful curricular experiences, and creating bridges between curricular and didactic development, reflective thinking and learning practices. In undertaking such analysis, the author draws on experiences from online courses that are part of distinct degree programs at the University of Guadalajara in Mexico.

Teaching and Training for Global Engineering: Perspectives on Culture and Professional Communication Practices,
First Edition. Edited by Kirk St.Amant and Madelyn Flammia.
© 2016 The Institute of Electrical and Electronics Engineers, Inc. Published 2016 by John Wiley & Sons, Inc.

6.1 Introduction

In the last three decades, diverse transformations in education have been challenging the nature of universities as we have known them. Among the most important factors affecting this situation are

- An unplanned expansion of student enrollment
- A systematical and deterministic institutional evaluation
- An increase in financial difficulties among educational institutions and difficult labor conditions of scholars
- A growing involvement of policy-makers and public decision-makers in the operations of universities
- Contradictions between quality of education and social equity across educational programs

Additionally, the increasing use of information and communication technologies (ICTs) in higher education has designed substantial modifications in how students are learning, how teachers are organizing different educational activities, and how both groups are experiencing the vast possibilities of virtual education. The potential force of ICTs can, however, amplify or diminish the use of a significant pedagogy. It can also decrease valuable participant social interaction depending on how ICTs are used to share information with and foster communication with students (e.g., student-to-student or student-to-instructor exchanges).

In many ways, these factors make the crisis of the contemporary university a part of a larger crisis of civilization [1] that involves all sectors of human activities. Such situations, moreover, are often seen as an extension of how the development of ICTs is transforming the functions, meanings and goals of universities worldwide. Understanding such factors requires educators, administrators, and students to rethink the substantive improvement of educational processes within the modern university. These changes affect the following processes among others:

- Curricular and didactics development
- Communication processes among teachers and students
- Institutional behavior regarding teachers' profiles and students' and teachers' performance (evaluation processes)
- Students' educational and professional competencies required for global job markets

Interestingly, the inclusion of ICTs within universities can be relatively seamless for pedagogical purposes if the integration of these technologies reproduces traditional face-to-face educational models. In fact, a participative and conscious pedagogical approach to ICTs can result in the creation of a flexible curriculum that can lead to wider student and teacher participation in online settings. Within this context, the instructional approach of autonomous learning can help educators maximize the benefits of using ICTs in educational environments, particularly in global educational contexts. Moreover, autonomous learning can help students not only to improve learning conditions but also to develop responsible practices as students and future participants in the global job markets.

> Educators and students using autonomous learning, based in curricular experiences, can improve pedagogical and communicational interactions for a meaningful construction of knowledge in higher education, particularly when such experiences are extended into global contexts.

This chapter examines how the strategic and pedagogical use of ICTs can better improve the curriculum and didactics development among teachers and students. The central thesis of the chapter affirms that educators and students using autonomous learning, based on curricular experiences, can improve pedagogical and communicational interactions for a meaningful construction of knowledge in higher education, particularly when such experiences are extended into intercultural global contexts (e.g., students located in different nations using ICTs to participate in the same online class).

To this end, the chapter offers a critical analysis of how autonomous learning can transform the use of ICTs and facilitate the development of more complex, diverse, inclusive and wider learning and teaching environments.

6.2 The Nature and Characteristics of Autonomous Learning

In general, knowledge allows social mobility, reaffirms identity and facilitates communicational interaction within human communities.

> Knowledge compels decisions and creates new contexts of action. Individuals are liberated from structures and they must redefine their situation of action under conditions of manufactured insecurity in forms and strategies of reflected modernization. [2, p. 25]

In this sense, knowledge is not simply a value or a currency within globalized markets, as many actual trends consider it to be [3–5]. Rather, it becomes a structural component of individuals' lives and the base of the process of human socialization. Communicational interactions, in turn, are the main vehicle through which knowledge can be created, modified, and shared with the members of a given group. And as online education can increase both the size and the scope of the group, it can foster new kinds of knowledge creation and dissemination.

The nature of learning, however, depends on the particular culture in which knowledge is designed and distributed. Learning is also connected to the practical actions of agents (i.e., the individuals who communicate/create and share knowledge) who belong to that culture. Consequently, online classrooms that bring together a wide swath of students and teachers from a range of cultures provide participants with unique and important knowledge creation and knowledge sharing experiences.

The more different the cultural groups interacting in such contexts are, the richer the kind and the scope of the knowledge that can emerge from such settings. Within such situations, the autonomy of learning emerges as a relevant alternative to link knowledge to social actions.

6.2.1 Historical Development of Autonomous Learning

Over the last 40 years, autonomous learning has evolved from schematic, behavioral, and cognitive models of learning to become a central component of a constructivist pedagogical model that focuses on the nature and quality of curricular experiences. Autonomous learning's first phase of development started in the 1970s and lasted until the early 1990s. The initial phase focused largely on the development of basic and complex cognitive skills (e.g., perception, description, induction, deduction, abduction and memory) as the construction of an object of study [6, 7]. This approach also relegated social and communicative interactions (e.g., effective communication, expert dialogue, and capacity to discuss and respect others' ideas, negotiation abilities) to a secondary or a background position.

Consequently, this cognitive approach relates curricular design and educational development to individuals' mental skills performance without taking into account the social interactions and processes of communication that made possible such performance. Although this approach opened spaces for curriculum development of an apparently deeper complexity, it also designed a considerable gap between didactics and curriculum that still exists in worldwide educational practices today.

It was only at the end of the 1990s that autonomous learning took on the role of an agent of engaged interaction and participation in more complex learning and communicational environments. What was particularly important about this shift is the fact that autonomous learning integrated a social constructivism approach. Further, with the new approach, autonomous learning went from a cognitive approach centered on the performance of mental skills to the examination of individuals' learning behaviors as constructed in relation to social interactions in an educational environment. Such social construction involved the following components:

- The inclusion of a paradigm of constructivism in pedagogical methods
- The idea of the curricular experience coming from constructivism
- The learning and teaching culture inside institutions of higher education

The development of these components forms the rest of the body of this chapter.

6.2.2 The Inclusion of a Paradigm of Constructivism in Pedagogical Methods

Constructivism is both a philosophical approach and a theory of learning. According to its tenants, knowledge design occurs as part of a process of creation and creativity from teachers and students rather than a process of acquiring knowledge through mental operations. Thus constructivism is related directly to experience in building an individual's own reality and own knowledge. Within this context, the three main principles of constructivism are as follows:

- The process of learning and teaching to build knowledge is not one of passive transmission and reception, but is rather an active one in which individuals are involved in acts of learning.
- Cognition requires continuous reflective thinking on how individuals learn and teach. This reflection process is called *meta-cognition* (i.e., learning how we learn something and the diverse conditions around the process of learning).
- Autonomous learning helps individuals to better adapt educational experiences in a way that makes them teach and learn in the most practical and meaningful manner.

The first principle (the "active" approach to learning) leads to more decentralized teaching activities, those that allow for more significant student participation and new student responsibilities in building and communicating knowledge. In this context, communicational interactions between teachers and students create a different pedagogical relationship where teachers no longer exert total control over the process of teaching and learning, while students take on more leadership in the design of knowledge within educational contexts. It is about a pedagogical pact—some authors use the term "contract" [8, 9]—where teachers and students share responsibility and autonomy.

To achieve this kind of educational context, autonomous learning draws upon particular communication and interaction approaches to guide interactions among teachers and students. Such approaches include

- The participation of both teachers and students as important constituents who construct part of the academic program
- The setting of objectives for student learning
- The selection of new and dynamic methods for sharing materials and information through uses of ICTs
- The deconstruction of the order of classroom sessions so that both teachers and students (as opposed to just teachers) have the right to present information and ideas

Within the pedagogical dynamic created by such approaches, it is also common to find the integration of different didactics, such as problem-based learning (PBL) and project learning applications (PLA).

The second constructivist principle—that of cognition—is related to the idea that, in autonomous learning situations, the only way students and teachers can effectively build knowledge is by being aware of how the involved parties teach and learn. This awareness is the origin of meta-cognition—a process that allows individuals to understand as as much as possible the how, why, what for, and under which conditions students and teachers build and communicate knowledge. According to this perspective, communication and social interactions between teachers and students should be more oriented to producing and to sharing knowledge rather than focusing on teachers establishing their social control of students or on student responses to such displays of power. Autonomous learning thus advocates a democratic approach to teaching and learning by setting different channels of communication interaction among teachers and students. This idea is one possible path for the construction of students' and teachers' autonomy in relation to the development of educational values.

> Autonomous learning helps individuals to better adapt educational experiences in a way that makes them teach and learn in the most practical and meaningful manner.

The last cognitivist principle points out the need to adapt the learning and teaching process to help students and teachers better interpret educational experiences. Doing so involves didactic devices that allow students and teachers to put together these experiences in whole units of meaning and sense.

> Once knowing is no longer understood as the search for an iconic representation of ontological reality, but, instead, as a search for fitting ways of behaving and thinking, the traditional problem disappears. Knowledge can now be seen as something that the organism builds up in the attempt to order the as such amorphous flow of experiences. [10, p. 46]

Important didactic devices used to achieve this end include pedagogical diaries, working portfolios, construction of learning objects [11], educational wikis and platforms (Moodle®, Wimba®), educational avatars (Second Life®), and communication interactions between teachers and students in Facebook® and Twitter®.

6.2.3 Autonomous Learning and New Approaches in Education

According to the autonomous learning perspective, the change in the nature of the interactions between students and teachers will no longer be ruled through acts of unilateral communication coming from the results of a particular test or exam. Now such education-focused interaction will be driven mostly by the meaning and sense of understanding an educational trajectory. By being aware of the educational trajectories of students, the whole educational community may be able to embrace an institutional pedagogic project centered on more meaningful and more conscious educational practices. It is this kind of widespread acknowledgement that is central to expanding educational approaches and contexts to a wider global base of students through ICTs.

For the past decade, some institutions of higher education in Canada, such as the University of British Columbia and University of Toronto, as well as in the Midwestern United States (the University of Kansas and the Ohio State University), and in Asia, such as Hong Kong, and in Central Europe, such as Open Universities in the Netherlands and Spain, have been implementing, diverse software to analyze students' educational trajectories. This is the case of SIM (Students Information Management), SIS (Students Information System), SML (Self management students Learning trajectories by Students), e-Portfolios, Wikiwijs (Wikiwise), etc. These examples indicate an increasing institutional interest in students' educational trajectories with a possible focus on future professional trajectories, societal needs, and job market requirements.

The following components within autonomous learning—improve communication and social interaction skills among teachers and students, and thus facilitate new pedagogical perspectives and practices in global contexts:

- Collaborative work
- Multidirectional communication
- Participation in curriculum and didactics design
- Active and expert participation
- Critical and autonomous thinking
- Knowledge comprehension based on educational trajectories
- Multi-evaluation processes based in understanding the educational trajectories (educational itineraries) of students and teachers

6.2.4 Curricular Experiences as a Result of a Constructivist Approach

As discussed, autonomous learning from a constructivist approach is based on rethinking and understanding the construction of knowledge from a diverse set of experiences. It also involves viewing learning and teaching from the perspective of meaning-making within an educational trajectory. Many of the scholars working in curriculum development and design related to action studies refer to this grouping of experiences and actions designed to build knowledge as *curricular experiences* [12–15].

According to Correa [16, 17], a curricular experience is one that involves sociocognitive reflection done in a way that commits individuals to certain situations—those that involve potential and accomplished acts of learning. Such acts, moreover, are achieved through a controlled narrative action. In order to access students' and teachers' curricular experiences, such experiences must be described, told, interpreted, understood, and placed coherently in an educational trajectory. In fact, curricular experiences can be mostly revealed through narrative processes. Learning has to be built upon *communicative interactions*—or by retelling the plot of these curricular experiences and then by sharing interpretations of what has taken place during the development of them.

Therefore, curricular experiences are always sparked by inquiry. They are about the desire to understand what something is, or how it works, or to help individuals to better understand what reality is, and what pushes individuals to learn about it. This process of inquiry is driven by questions, for example, "What is it?" or "How does it work?" or "Why does it react this or that way?" There is, moreover, a key difference between the uses of questioning in a traditional learning approach and in an autonomous learning approach. In an autonomous learning approach, questions are oriented toward the understanding of heterogeneous curricular experiences and the way they were or are embodied in an educational trajectory. This is opposed to a traditional learning approach in which asking questions represent a strategy to collect concepts in order to understand a particular phenomenon/situation.

6.3 Understanding and Applying Autonomous Learning

To better conceptualize the ideas presented thus far, consider the following context: We can say that when questioning students about the meaning of temperature from a traditional approach, it is possible to recover the concepts of "grades of temperature." Such grades might, in turn, define basic cognitive categories such as "hot or cold" (e.g., "hot" is defined according to a certain number of Celsius or Fahrenheit degrees"). An autonomous learning approach, by contrast, will lead students to recover a series of curricular events that made it possible to experience "hot and cold" by themselves (e.g., "Remember the time when it was about to snow last winter, that was really cold, it was minus 8 Celsius degrees.").

In understanding this experience, students will figure out that, to put the concept of temperature in their educational trajectory, they must express and recover (in a narrative process) that idea in terms of more cognitive resources than in terms of "grades of temperature." (It is one thing to say that 110°F is hot. It is a completely different thing to say, "I remember being outside in 110°F weather; that was hot.") For instance, in affirming that, "When I was in the mountains during the summer, I could realize that it was cold, even though it was not winter time." This personalization of the concepts allows students to better understand how the curricular experience involves students in a particular and collective undertaking while learning about a particular subject.

But interestingly, in using autonomous learning, we learn not only through direct experience, but also by using the concepts of mediation, reference, and situated learning to understand reality. Thanks to technology and our uses of it, we have learned that there are planets and moons without being actually in space. As well, we understand things through reading and research. History and fiction can show things that we do not experience directly. Retelling historical events and using figures of speech as metaphors, similes, and analogies can help us to understand reality. Also, we understand through situated learning that it is within a particular context that we understand reality, when we are confronted with a particular problem and even more when we project ourselves in actions.

6.3.1 Using Problem-Based Learning within Autonomous Learning

PBL helps clarify how notions can become concepts, theories, and ideas. PBL requires inquiry, communication, and developing critical thinking. Such clarity is important to developing the understanding needed to address such a complex problem.

Using a PBL approach to this end often involves presenting, reviewing, and solving a given problem according to the following steps:

Step 1: Look for notions in a story (normally in terms of a problem)

Step 2: Build concepts

Step 3: Prioritize concepts

Step 4: Build learning objectives

Step 5: Delve deeper within the knowledge of these concepts through direct data or theoretical sources

Step 6: Build semantic networks that make sense through didactic tools (synoptic frameworks, conceptual maps, etc.)

Step 7: Present a plenary with other students to discuss these semantic networks.

Step 8: Make inferences and conclusions

By using these steps, students can create diverse graphic representations of a given situation. Such representations can help students to use the narrative process to better understand reality and the nature of a particular problem. Let's analyze the following news story published in the UK-based journal, *The Guardian* [18].

> Unusually hot summers, and the destructive droughts and wildfires that follow, are the product of climate change, according to a study of recorded global temperature data by a prominent NASA scientist.
>
> The study uses recorded temperature data, rather than prediction models, to assert that climate change is responsible for recent extreme weather events including last year's droughts in Texas and Oklahoma, the Russian heat wave in 2010 and the European heat wave in 2003.
>
> The author of the study and head of NASA's Goddard Institute for Space Studies, Dr. James Hansen, said: "We now know that the chances these extreme weather events would have happened naturally – without climate change – is negligible." Statistical data in the study shows that temperature extremes are becoming more frequent and more intense across the globe. This does not eliminate the possibility of cooler-than-average summers in the future, but it does mean that the probability of unusually warm summers has greatly increased, the report said.
>
> By comparing temperature data from the past 30 years to the 30 years prior, Hansen and his colleagues say the data shows that temperature extremes have increased from affecting less than 1% of global land area to an estimated 10%. Hansen explained that increased greenhouse gas emissions "load the dice," or make the occurrence of these extreme temperature events much more likely.
>
> Hansen said, in a press conference that to combat the change there needs to be a global effort to phase out dependence on fossil fuels in the long run. Hansen said:

> "Frankly, it would not be that difficult to do it if we put an honest price on fossil fuels," and went on to advocate for something similar to a carbon tax where companies must pay a fee to account for their use of fossil fuels.

Now, if we wished to use this item in an autonomous learning context, we would undertake the following steps:

Step 1: Choose a story to narrate as a problem. We use the example of *The Guardian*.

Steps 2 and 3: Build concepts and prioritize concepts: In our example: *Main concepts:* Climate changes, temperature, greenhouse gas emissions, fossil fuels. *Complementary concepts:* hot, warm, cool, and cold.

Step 4: Build learning objectives. Students should investigate the concepts they are working with, normally at a library and in online databases. But in doing so, it is important to first determine what we have to learn about these concepts by building some learning objectives, for instance: (a) analyze the definitions of climate changes and how they affect the course of nature; (b) identify the materials that contaminate the environment the most and how the overall process of contamination occurs; and (c) analyze the implications of the contamination through gas emissions.

Step 5: Delve deeper into the knowledge of these concepts through direct data or theoretical sources.

Step 6: Build semantic networks that make sense of the concepts through didactic tools (synoptic frameworks, conceptual maps, etc.). These graphics, moreover, can help students convert notions to concepts. In Figure 6.1, we can see the integration of concepts relationships. However, the networks can become more and more complex as we learn more about the problem.

Step 7: Present a plenary with other students to discuss these semantic networks. Normally the plenary is composed of one or two one and a half-hour sessions. It can be scheduled according to the needs of the group and their learning rhythms.

Step 8: Make inferences and conclusions. In this section the teacher participates actively in order to find the possible logic and to be aware of the different educational processes the students go through.

As this example helps to illustrate, teacher–student interactions become very dynamic because they are not related directly to specific content but mostly to a problem. Autonomous learning uses different didactics to build a richer learning environment and then to design meaningful curricular experiences such as PBL, PLA, learning objects, and narrative devices (such as fables, tales, and stories). However, the importance of the curricular experience remains in helping students develop meaningful communications and interactions with one another and with their teachers.

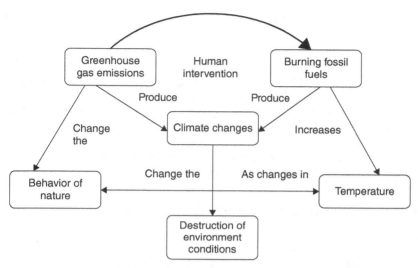

FIGURE 6.1. Integration of Concepts Relationships [16]

6.3.2 Connections to Curricular Experiences

Curricular experiences happen throughout the learning-teaching processes and they are recovered from narrative processes (biographical methods, storytelling, and narrative inquiry). We can confirm the quality of these experiences and the pedagogical conditions that make them meaningful for students and teachers. While talking about the theory of action research, Elliot [19, p. 10] acknowledged that

> In order that teachers and students achieve educationally worthwhile changes in curriculum and pedagogy, action research needs to address concerns about the educational quality of students' curricular experiences, and the pedagogical conditions under which they are accessed.

Through curricular studies Mckernan [20] focuses his analysis not on the performance of students, but on improving the quality of human actions among educational communities based on theories of social action. Such a perspective helps us understand that a curricular experience is considered meaningful (and valuable) when students and teachers understand knowledge as a process of learning construction that is embodied in an educational trajectory.

Consequently, learning and teaching communication processes are directly connected with narrative educational analysis regarding the quality of curricular experiences. Within this disciplinary domain, it is possible to integrate the ideas of Pineau and Le Grand [21]. Both authors see a close relationship between biography and education (*biographie et formation*), emphasizing that all communication and social interaction can be used for a pedagogical purpose and all involve stories of education (*récits de formation*):

> The actor becomes an individual that speaks from his/her experience, generally, with a meaningful purpose. This is the birth of an interlocutor individual in the twofold sense of access to the speech and dialogue, but also in the sense of amazement and surprise. It is the emergence of a personal world in genesis, in formation. [22, p. 10]

Thus, storytelling within a narrative process can help teachers to understand how students learn and the problems teachers and students have in teaching and learning.

6.3.3 Curricular Experiences and Educational Stories

Ricoeur [23] [24] [25] notes that, as humans, we not only tell stories, but we actually *are* our stories. According to this idea, educational stories relate the most basic and meaningful aspects of the educational trajectories (itineraries) and do so in a way that includes the live experiences of students and teachers. Pineau and Le Grand [26, p. 10], affirm this idea when writing:

> Life stories, interpret as the construction of sense from temporary personal facts, are imposed since life is imposed. There are no stories imposed therein. Life is full of stories about problems.

Yet, it is not enough to describe and analyze the collected stories of an individual for educational purposes. Rather, there must be a confrontation with reality and conciseness. At this point, educational stories (*récit de formation*) are closely related with the term *Bildungsroman*—an idea that refers to the capacity of individuals to reflect deeply on their actions by retelling them through a meaningful story or group of stories.

As Rorty [27] explains, through the concept of edification, the *Bildungsroman* approach makes it possible to recognize the construction of knowledge from narrative processes (i.e., as a communicative way to socialize). McDowell [28, p. 84] further defines *Bildung* as "the process of becoming capable of playing the game of giving and asking for reasons," or of "having one's eyes opened to reasons [...] having one's eyes opened to reasons at large by acquiring a second nature." As for the educational domain, we can define *Bildung* as the capacity students and teachers have to reflect on their curricular experiences and to share them in order to improve their communicational interactions and to embody the comprehension of these experiences in their educational trajectory. The overall purpose of *Bildungsroman* [29] does not stop with constructing knowledge, but goes on to help students achieve consciousness of reality and self-realization, offering autonomy in the processes of learning.

Autonomous learning becomes a philosophy of education in the sense that the autonomy comes from the awareness of students' and teachers' educational trajectories and the possibilities of learning, mobilizing knowledge, and acting in a particular educational community. Hence, working within autonomous learning, students, teachers, and administrators have the responsibility to improve the communicational interactions between them and toward society. Transforming curricular relations is not about improving student performance. Rather, it is about providing students with quality experiences that result in meaningful learning. Creating such experiences involves thinking of students, teachers, and administrators as part of a community in which the members are in constant communication with each other.

This autonomy during the process of learning designs a flexible curriculum, since teachers do not exclusively control the curricular dispositions and the contents included in a course or seminar, neither do educational administrators, nor do the students. It is more of a participative movement of conflict and freedom [30], where all members of the educational community must participate, producing a networking curriculum. Within this framework, the "Aha!" moment means that comprehension comes from reflecting deeply about an overall educational trajectory, where concepts, categories, notions, ideas, emotions, intuitions, and defaults are around the construction of curricular experiences.

6.4 The Role of ICTs in Autonomous Learning

ICTs offer multiple technical and educational options for creating flexible teaching and learning activities. (Of course, such opportunities are generally connected to the costs associated with procuring such technologies and maintaining the conditions in which they are used.) Moreover, important educational and social issues have arisen around the development of ICTs, particularly as they relate to education. Among the issues educators must now consider in relation to ICT-based or ICT-driven education are processes of communication, participation, socialization, as well as autonomy and learning.

It is crucial that different pedagogical processes accompany ICTs and facilitate curricular flexibility and reflexivity in order to help students engage in significant and communicative learning acts. If such factors are recognized and realized, then ICTs can contribute significantly to building curricular experiences that enhance cultural, cognitive, and social interactions. Moreover, ICTs can be used to foster and maintain rich interactions throughout the whole processes of teaching and learning. ICTs can create a collaborative educational environment engaging students and teachers in meaningful collaborative learning through peer evaluation in an intercultural global scale.

The possible benefits of using ICTs to apply autonomous learning within international contexts include the following actions:

- Deregulation of the space and time of learning. ICTs expand the length of curricular experiences. They are not restricted to real-time schedules in classrooms, or to a limited learning community.
- Deinstitutionalization of learning spaces. Their use is not limited to classrooms, but could inhabit different contexts: school, home, street, or place of work.

- Development of curricular experiences in a variety of languages among virtual communities, varied cultures and even different educational traditions.
- Construction of local and global learning communities of a synchronic and diachronic nature.
- Allowing of students and teachers to interact with multicultural and multilingual educational communities.

Because it represents a communicative and a social basis where it is possible to build networks of interaction within narrative learning processes, the autonomy of learning—particularly on a global scale—can be improved by using ICTs.

6.5 The Culture of Autonomous Learning Inside Institutions of Higher Education

Learning in autonomy is not limited to a particular methodology of teaching. Instead, this process requires a series of basic conditions that can enable students and teachers to develop reflective and argumentative thinking, communication skills, meaningful interactions (like the capacity to develop critical judgment and to convey and share ideas), and pedagogical awareness (to realize how, why, and where knowledge is designed and shared). These conditions include

- Building a flexible curriculum allowing students and teachers to develop their own educational trajectories according to their conditions and interests
- Transforming didactics by harmonizing educational technology innovations and contemporary pedagogical practices
- Rethinking evaluation and critical didactics by situating the attention in the students and teachers educational trajectories
- Transferring more academic responsibilities to students so that students and teachers can build and adapt the whole process of teaching
- Transforming administrators and policy makers of higher education's model of management of education to curriculum leadership, as a responsibility of the entire educational community by placing in the core of the reforms the educational and societal project

6.5.1 The Culture of Using ICTs to Build and Develop Curricular Experiences

By using ICTs, students and teachers can develop digital literacy and adopt communicative interactional process not only as skills, but also as a culture of learning in intercultural communities. When using ICTs to engage in educational activities, it is compulsory to account for the cultural, social, and economic origins of students and teachers as part of the educational, social, and professional trajectories of individuals involved in a given educational context. In Latin America especially, technology has not reached all communities, and also, the populations that do have access to ICTs can experience differences in the quality of access and the type of connectivity they have.

The development of these trajectories (social, educational, and professional) will define the personal and collective identities of those using ICTs to interact in educational settings. To communicate effectively in intercultural and autonomous learning communities—particularly online ones—students and teachers alike must not only develop certain cognitive skills, but they must also develop a culture of recognition [31, 32] that allows and fosters diversity of identities in such contexts. This educational culture is not focused on controlling students' interactions or student–teacher communication. Similarly, it is not about building one hegemonic identity or language through ICTs or restraining students to the schematic program of a course or a seminar. Rather, ICTs may allow teachers and students to create better educational environments and good quality curricular experiences.

Critical didactics implies the intelligent use of different educational strategies to make sense of and give meaning to a curricular experience. For instance, in spite of the fact that a conceptual map is a strong didactic strategy, it won't work for a general description. It will be more accurate for students to use a synoptic framework to describe and to classify elements in a hierarchic order.

The use of ICTs in developing PBL may promote the construction of more meaningful curricular experiences, taking into account that PBL requires a continued communicational interaction between students and teachers, and that ICTs are able to offer a good connective platform for a wider educational interaction. Finally, the use of ICTs and PBL can improve relations between teachers and educational administrators by giving them the opportunity to work together to design meaningful qualitative educational indicators.

6.5.2 Process of Evaluation within Educational Trajectories

Evaluation is a process of validating the actions individuals undertake in learning situations. Evaluation, however, is not exclusively confined to assessing cognitive abilities. Rather, according to Alvarez [33, p. 21],

> Paradoxically, evaluation has to do with activities of qualifying, measuring, correcting, classifying, certifying, reviewing, and passing tests, but evaluation cannot be confused with all of them. They share a semantic field, but they differ by the resources that they use and the uses and purposes, which they serve. These are activities that play a functional and instrumental role. From these instrumental activities it is not possible to learn. Evaluation transcends them. Just where they fail, it starts the educational evaluation. To have a real evaluation it is needed the presence of individuals.

Evaluation requires communicational interactions among teachers and students in order to concentrate the evaluative effort in a way that leads to formative actions. Evaluation as valuation of individuals' actions is, in turn, related to reflection. So, to be meaningful, evaluation activities should be consistent with the interpretation and valuation of curricular experiences involved in the learning and teaching processes.

Therefore traditional evaluation practices differ from the process of evaluation within autonomous learning in the sense that the first is concentrated in students' performance and the second on the quality of curricular experiences students and teachers have. For instance, if a curricular experience is focused on memorizing and understanding concepts because the proximity of an exam, students might have difficulties in relating these concepts to their educational trajectory. However, if these same students have the opportunity to decide together which concepts they need to learn and to discuss them with each other and with the teacher. I so doing, they will not focus on their performance on a given exam, but on the quality of the interactions in this session. Evaluation will change from an instrumental activity to a dialectic process.

The following testimonies from students of the University of Guadalajara, in Mexico, will shed some light on the ideas we have mentioned in this last section. To begin, one student noted,

> As students we always have worries about how are we going to be evaluated by teachers. Some teachers evaluate concepts that we have to learn by heart, and they are not interested about our process of learning, the problems we had or have in learning these concepts. So we just repeat and repeat many times these concepts until we know them all by heart. I think it is very boring and we easily forget a concept. We do not have the opportunity to relate this concept with others and sometimes evaluation is measuring the capacity of memory and not if we understand or not a concept. Teachers must teach us how to improve learning not only memory. [34]

This testimony shows a disconnection between a particular experience (learning concepts by heart) and the process of evaluation. This type of exam is used only to measure the students' ability to memorize based on remembering a collection of concepts by heart. In this traditional model of evaluation, the exam restricts the design of knowledge by describing concepts by heart to the teacher at the moment of the exam. Therefore, the exam represents an arbitrary and a specific action from the teacher to the student. In such situation, students work to win the teacher's validation/approval than to determine if they, themselves, have actually learned anything. On the contrary, a coherent curricular experience always provides (or requires) a learning (educational) story and a network of relationships between concepts and ideas that have to be shared with other students in order to build knowledge.

6.5.3 Autonomous Learning, ICTs, and Knowledge Construction

Autonomous learning fosters communicational interactions between students and teachers by encouraging them to discuss the process of building knowledge. A concept needs to be discussed among students and teachers in order to design an educational story. In this sense, evaluation gets more complex by becoming self-evaluation (as a personal reflection on the conditions when a person had learned something or about something), co-evaluation (students evaluating the learning processes of other students), and hetero-evaluation (students sharing their experience with other persons who did not participate in the beginning of the process of learning). Consequently, evaluation is an important component of the learning processes. As one student explained,

> I'm studying a degree program in Learning Technologies, and we have this course called *Learning Theories*, using PBL, and evaluation is something we do ourselves every time we get together in class, but it is not the traditional test, not an exam, it is just the idea of asking oneself about the process of learning. Especially when the teacher told us: ask yourself if you can talk about a topic, and how much can you do it or about an important problem on these days. It is difficult because we know things by heart but to put these concepts in reality is quite difficult. At the moment that we discuss with classmates it is so hard, sometimes I thought that I really knew about the topic we are discussing but other student says something I haven't consider before, so even I get confuse sometimes, I realize that the opinion of others help me to understand better. Sometimes I don't think that a traditional exam is necessary at all when we are using PBL. [35]

The overarching idea is that evaluation moves from an arbitrary activity to an aware and intentional educational praxis among students and teachers. Through PBL, the communicational interaction is enriched with different students and teachers' interactions during the process of teaching and learning. More importantly, this approach allows for meaningful knowledge creation because students and teachers can discuss ideas and situate this discussion in the context of a particular problem.

In general, using traditional methods of evaluation means the communicational interactions of students and teachers are defined by the type of evaluation teachers do. Many times students critique teachers, not because of the way they teach, but because of the way they evaluate students, "Tell me how the teacher evaluates and I will tell you how he/she teaches" is a phrase that is commonly heard around students. Here, it is true that evaluation is an important part of the learning process. However, it seems that the learning and teaching processes are fragmented because of a traditional approach that makes learning appear as a meaningless activity. Fernández [36, p. 42] affirms these ideas when noting:

> The practice of evaluation can be explained by the way in which the functions are carried out by the school and therefore its realization is conditioned by numerous personal, social and institutional aspects; at the same time, it has an impact on all the other elements involved in schooling: transmission of knowledge, relations between teachers and students, interactions in the group, methods practiced, discipline, expectations of students, teachers and parents, individuals' valuation in society, etc. Evaluation helps decisively, therefore, to configure the educational environment.

The process of evaluation through autonomous learning and based on PBL allows students and teachers to set learning objectives together (i.e., to learn concepts that allow students and teachers to discuss them with some sense of epistemic deepness). Even when working in groups, students have to choose what concepts they are going to learn in order to participate in the plenary session.

Getting back to the analysis of the problem of climate change, students might have to build a learning objective to understand the diversity of the concept of "pollution." It is a central concept, but probably some of the students might find other concepts they need to review (e.g., fossil fuels) in order to have a better understanding of the dynamics of climate change. Therefore, continued evaluation based upon PBL is not just a technique to teach or learn. Rather, this approach implies the recognition of what actions are needed to build knowledge and comprehension.

6.5.4 Rethinking Evaluative Contexts and Approaches

Within autonomous learning, each curricular experience involves diverse evaluative acts. The reflections that motivate these acts are not isolated from the intention of acts of learning. For instance, activities such as asking, evaluating options, arguing, consulting, reformulating, planning concepts to learn, building examples, and designing problems are all evaluation components that enrich the processes of learning and teaching. However, not all students are prepared to link curricular experience and acts of evaluation. As one student explained,

> I get confused when there are no exams, and then I ask the teacher, when is it going to be an exam of this unit? And once she answered me: why do you want to have an exam? My answer was, to have a grade. I realized from her eyes that grades were not a priority in our course, so I was surprised. I don't know, we always expect exams, is like the serious part of the courses, it is that the other part of the course is that important, but when reaching the exams the course got serious. Using PBL I don't know how to react, it is very difficult for me and, I thought it was a good idea because a friend of mine told me that it would be nice to try something that is new at the university, I mean PBL. But even I think it is very interesting but too much work [sic], I will get back to the traditional system, it is easier. [37]

Generally, students wait for the exam in order to finish with a topic or content, such as unit one, unit two, etc. In autonomous learning through PBL or PLA, it is important to tie up all the knowledge students have when facing a problem. However, taking into account that teachers probably would like to maintain traditional exams, as a very effective way to control students, it will be very difficult for students to participate in building knowledge or educational trajectories. Accordingly, autonomous learning becomes a stream of curricular experiences not governed by the exams, but by the quality of curricular experiences. Thus, in using autonomous learning, evaluation represents a bridge between one type of curricular experience and another.

Curricular experiences are the substrate of a culture of learning and a path to learning in autonomy. However, this autonomy does not intend to abandon students to their fate. In most cases, it gives students the opportunity to express themselves, to gain a better understanding of what is taking place, and to act ethically. Learning in autonomy promotes an intercultural atmosphere and enriches communication for the culture of the class is not imposed by the teacher; on the contrary, the dynamic of the encounter with different cultures is what rules the sessions. As one student put it,

> We have this course and it is delivering on line and in English [sic], but we have some face-to-face sessions, and we have three international students with us, and two other from Dominican Republic that follows the course online. It is very interesting, because they help us to improve our English skills, but at the same time is difficult because we cannot communicate with them through Spanish, so we are compelled to speak in English and we have to make a big effort. It's like an international culture within our course and it isn't a simulation, like in other courses at our university here in Mexico, where we speak in English and we pretend to be in an international atmosphere, but all of us speak Spanish as a first language. We don't take other courses seriously because we all know that in any moment we can say things in Spanish if someone doesn't understand in English. I have been in Face (Facebook) the other day, and I found these classmates from abroad in our course like this guy from California, and now we are friends. I think it's very good, because we learn and we get friends at the same time. [38]

The possibilities of using ICTs increase global participation and the more the students interact with other students with different cultural backgrounds and origins, the broader and richer the resulting design of knowledge will be.

Finally, evaluation is a very important indicator of how the educational trajectory is being developed through a period of time. By looking at the different virtual educational strategies such as portfolios and wikis, among others, students and teachers can see their overall participation in the process of constructing knowledge. At the same time, it is a golden opportunity to improve students' and teachers' communicative interaction through chats and e-mails, SMS, and talking about the process of learning and teaching. Thus, the nature of communication between students and teachers must change from institutional control to social responsibilities. Doing so can facilitate the best possibilities to learn, to expand the curricular experiences, and to understand more about students' and teachers' educational trajectories.

6.6 Conclusion

Autonomous learning offers communicative spaces of interaction that extend and enrich curricular experiences. It is through narrative processes that autonomous learning can better express its educational possibilities to build knowledge. Through autonomous learning, students and teachers from institutions of higher education are compelled to engage in communicative interactions that allow them to generate a new culture of learning and teaching based on symmetrical relations. This is not about placing students in the center of processes of learning. It is not about de-centering teachers. Rather, it focuses on the meaning and sense of curricular experiences in an environment that encourages communication.

The process of learning is enriched thanks to a social interaction focus on comprehension and socialization. Learning becomes autonomous to some extent when it does not respond to the contents and the rhythms of teachers, programs, or administrative agendas and interests. It must instead involve creating learning communities that focus their interests on building and offering quality curricular experiences and on understanding the result of them as part of the students' educational trajectories. ICTs, in turn, represent a privileged non-regulated space and time structure for rich and complex communicative interactions among students and teachers. Such situations, moreover, offer unique possibilities for international and intercultural participation. Such factors could be used to encourage students and teachers to face the construction of three structural components among autonomous learning: a new culture of learning, more complex processes of comprehension, and dynamic processes of socialization.

For these reasons, educators in Mexico need to consider the importance of linking curricular experiences to autonomous learning mediated by ICTs in order to improve evaluation processes and the overall process of teaching and learning. Moreover, they must do so with a positive and open attitude, and use conciseness to reflect deeply on the implications of the newly acquired knowledge. Institutions of higher education, in turn, require a pedagogical culture that emphasizes and prioritizes the development of students, teachers, and administrators through a critical pact and an ethical intention toward significant processes of teaching and learning.

> Learning becomes autonomous to some extent when it does not respond to the contents and the rhythms of teachers, programs, or administrative agendas and interests. It must instead involve creating learning communities that focus their interests on building and offering quality curricular experiences and to understanding the result of them as part of the students' educational trajectories.

The administration of institutions of higher education in Mexico must provide help to develop an intelligent ICTs infrastructure and the pedagogical conditions that allow ICTs to activate continuous communication between students and teachers. This communications must enhance narrative analysis to interpret curricular experiences and promote a wider interaction between students and teachers. Also, it is important for all undergraduate and graduate students in the process of building their educational trajectories (and facing future employment opportunities and obstacles) to adopt a culture of autonomous learning in order to develop the critical thinking skills needed to use concepts, ideas, and theories properly so they can communicate coherently, critically and contextually. But more importantly, autonomous learning can develop a profound clarity of social reality and help students better understand the possibility of social agency.

References

1. E. Morin, *La Voie*, Paris, France: Fayard, 2011.
2. U. Beck, *World at Risk*, Cambridge, MA: Polity Press, 2000.
3. P. Druker, *Post-Capitalist Society*, London, England: Butterworth/Heinemann, 1993.
4. M. Young, *Bringing Knowledge Back In*, London, England: Routledge/Falmer, 2007.
5. P. Brown, H. Lauder, and D. Ashton, "Towards a high skills economy: Higher education and the new realities of global capitalism," in *World Yearbook of Education 2006, Geographies of Knowledge, Geometries of Power. Framing the Future of Higher Education*, D. Epstein, R. Boden, R. Deem, F. Rizvi, and S. Wright, Eds. London, England: Routledge, 2008, pp. 190–210.
6. A. L. Brown and A. S. Palincsar, "Introducing strategies learning from text by means of informed, self-control training," *Topics in Learning and Research Disabilities*, vol. 1, pp. 1–17, 1982.
7. I. Mitchell and A. Swarbrick, *Developing Skills for Independent Learning*. London, England: CILT, 1994.
8. A. Font, "La incidencia del aprendizaje basado en problemas (ABP) en la integración laboral de los licenciados en derecho," *Revista de Derecho Mercantil*, vol. 3, no. 287, pp. 237–284, 2013.
9. J. Rue, *Aprendizaje Autónomo en Educación Superior*. Madrid, Spain: Narcea, S. A. de Ediciones, 2009.
10. E. Von Glasersfeld, "The reluctance to change a way of thinking," *The Irish Journal of Psychology*, vol. 9, no. 1, p. 46, 1984.
11. D. A. Wiley, *The Instructional Use of Learning Objects*, Bloomington, IN.: Agency for Instructional Technology, 2002. Available at: http://www.ltimagazine.com/ltimagazine/article/articleDetail.jsp?id=5043 (accessed October 14, 2012).
12. J. Elliot, *La Investigación-Acción en Educación*, Madrid, Spain: Ed. Morata, 1999.
13. J. Mckernan, *Investigación y currículo*, Madrid, Spain: Ed. Morata, 2000.
14. J. M. Álvarez Méndez, *Entender la Didáctica, Entender el Currículo*, Madrid, Spain: Editorial Miño y Dávila, 2001.
15. C. Correa Arias, "El gesto evaluativo: Los retos del aprendizaje basado en problemas en el marco de la educación superior contemporánea," in *Aprendizaje Basado en Problemas en la Educación Superior*, vol. 1, C. Correa Arias and J. A. Rúa Vasquez, Eds. Medellín, Colombia: Editorial Universidad de Medellín, 2009, pp. 195–213.
16. C. Correa Arias, "Didáctica del aprendizaje autónomo en la formación de investigadores en ciencias sociales," in *Aprendizaje Autónomo en la Educación Superior*, vol. 2., C. Correa Arias and J. A. Rúa Vásquez, Eds. Medellín, Colombia: Editorial Universidad de Medellín, 2012, pp. 78–99.
17. C. Correa Arias, "Autonomous learning as a supporter of curricular experiences: The significance of PBL in online education," *Educational Studies*, vol. 17, pp. 364–370, 2011.
18. A. Holpuch, "Greenhouse gas emissions, fossil fuels," *The Guardian*, July 9, 2012.
19. J. Elliot, *El cambio Educativo desde la Investigación Acción*, Madrid, Spain: Ed. Morata, 1996, p. 10.
20. J. Mckernan, *Investigación y Currículo*, Madrid, Spain: Ed. Morata, 2000.
21. J. Pineau and J.L. Legrand, *Les Histoires de Vie*, Paris, France: PUF, 2007.

22. J. Pineau and J.L. Legrand, *Les Histoires de Vie*, Paris, France: PUF, 2007, p. 10.
23. P. Ricoeur, *Temps et Récit. Tome I: L'intrigue et le Récit Historique*, Paris, France: Le Seuil, 1983.
24. P. Ricoeur, *Du Texte à L'Action. Essais D'Herméneutique II*, Paris, France: Le Seuil, 1986.
25. P. Ricoeur, *Soi Même, Comme un Autre*, Paris, France: Le Seuil, 1990.
26. J. Pineau and J.L. Legrand, *Les Histoires de Vie*, Paris, France: PUF, 2007, p. 58.
27. R. Rorty, *La Filosofía y el Espejo de la Naturaleza*, Madrid, Spain: Cátedra, 2001, p. 84.
28. J. H. McDowell, *Mind and World: With a New Introduction*, Cambridge/ London: Harvard University Press, 1996.
29. K. Stojanov, "The concept of Bildung and its moral implications," *presented at the Annual Meeting of the Philosophy of Education Society of Great Britain*, New College, Oxford, England, 2012.
30. P. Ricoeur, "Rebâtir l'Université," *Le Monde*, June, 1968, p. 1.
31. A. Honneth, *The Struggle for Recognition*. Cambridge, MA: MIT Press, 1996.
32. P. Ricoeur, *Parcours de la Reconnaissance. Trois études*. Paris, France: Folio, 2005.
33. J. M. Álvarez Méndez, *Evaluar para Conocer, Examinar para Eexcluir*. Madrid, Spain: Morata, 2001, p. 21.
34. Student 1, Personal Interview (Interviewee Name Kept Confidential), University of Guadalajara, 2010.
35. Student 2, Personal Interview (Interviewee Name Kept Confidential), University of Guadalajara, 2010.
36. L. Fernández, *Instituciones Educativas*, Buenos Aires, Argentina: Ed. Paidos, 1998, p. 42.
37. Student 3, Personal Interview (Interviewee Name Kept Confidential), University of Guadalajara, 2010.
38. Student 4, Personal Interview (Interviewee Name Kept Confidential), University of Guadalajara, 2010.

7

E-Learning and Technical Communication for International Audiences

Darina M. Slattery
University of Limerick

Yvonne Cleary
University of Limerick

This chapter examines the pedagogical strategies adopted by faculty teaching in the Master of Arts in Technical Communication and E-Learning program at the University of Limerick, Ireland. These strategies enable students to design and develop global technical communication and e-learning content. By the end of the program, students have participated in virtual team projects with students in the United States and have used a variety of media to write and design localizable content (e.g., brochures, web sites, and e-learning courses, for international audiences. The chapter discusses the literature that informs pedagogical approaches used in this MA program and then outlines the program, its components, and its learning outcomes).

7.1 Teaching Technical Communication and E-Learning: An Introduction

This chapter describes the pedagogical strategies adopted by faculty teaching in the MA in Technical Communication and E-Learning program at the University of

---

Teaching and Training for Global Engineering: Perspectives on Culture and Professional Communication Practices,
First Edition. Edited by Kirk St.Amant and Madelyn Flammia.
© 2016 The Institute of Electrical and Electronics Engineers, Inc. Published 2016 by John Wiley & Sons, Inc.

Limerick, Ireland. The purpose of this chapter is twofold: First, it will inform teachers of the importance of teaching learning pedagogies and intercultural communication to technical students. (It is our students who will later be required to design and develop global technical communication and e-learning content when they enter the workplace). Second, the chapter will provide teachers with ideas and approaches for how to teach these subjects more effectively by providing insights and highlighting sample assignments that we have successfully integrated into our curriculum. As the majority of our graduates successfully secure technical roles in e-learning and technical communication companies, we believe our approach will be of interest to other practitioners.

In Sections 7.2 and 7.3, we present an overview of the key literature relating to learning pedagogies. In examining these topics, we also discuss why we believe teachers should consider incorporating such approaches into the professional and technical communication curriculum. Our belief is that knowledge of these approaches not only informs students of key strategies they may need to integrate into e-learning courses in future, it also informs teachers' current pedagogical strategies.

In examining these items, we also overview key literature relating to teaching intercultural communication and focus on three particular areas—collaboration, information design, and rhetorical awareness. (These areas represent important curriculum focal points for future technical communicators and instructional designers, who must often work in multicultural teams, and design content that may be accessed by international audiences.) We also discuss the Irish context, with a particular focus on the technical communication, e-learning, and localization industries in Ireland. This latter section highlights the relevance of our curriculum to the workforce.

In Sections 7.5.1 and 7.5.2, we provide an overview of the MA program in Technical Communication and E-Learning program at the University of Limerick. In so doing, we discuss courses that comprise localization and/or intercultural communication components. In Section 7.6, we focus on three main types of assignment—a collaborative (virtual team) assignment, a design and development (web site) assignment, and a sample of writing assignments. (We selected this focus because the assignments exemplify collaboration, information design, and rhetorical awareness challenges for students.) In discussing each assignment, we outline the objectives of each and the evaluation criteria we use to assess student work. We also discuss the relevance of each assignment to the workplace.

Finally, we conclude in Section 7.8 with a summary of how our chapter contributes to the existing literature on teaching intercultural communication. Yu [1] argues that while the technical and professional communication field has done an effective job of integrating international issues into the curriculum, the question of assessing intercultural competence remains problematic. We agree that there is a dearth of research that examines the process, challenges and especially the outcomes of teaching students how to develop technical communication and e-learning products that can cross cultures effectively. This chapter represents one attempt to address this research gap.

7.2 An Overview of Learning Pedagogies

We teach learning pedagogy on the program for two main reasons. First, instructional designers and e-learning content developers need an in-depth understanding of their audience of learners, so they must study how people learn. Second, they need to understand how different teaching strategies may impact learners' achievement of learning outcomes. Our students also need to study intercultural communication, as they will increasingly be required to develop content for international audiences. We will examine the key literature on three main approaches to instructional design for e-learning: behaviorism, cognitivism, and constructivism. This section also examines newer pedagogical strategies that are often used in conjunction with all three approaches. We will then look at the key literature related to the teaching of intercultural communication and do so with a focus on three key areas: collaboration, information design, and rhetorical awareness.

The disciplines of E-learning and of technical and professional communication are two closely-related fields. The content in e-learning courses is typically written by instructional designers, so it is essential that they have knowledge of pedagogical theories. As Ertmer and Newby note, "Learning theories provide instructional designers with verified instructional strategies and techniques for facilitating learning, as well as a foundation for intelligent strategy selection" [2, p. 1].

Within the context of instructional design, instruction has been defined as "a set of events that affect learners in such a way that learning is facilitated" [3, p. 3]. Instructional design, in turn, involves the systematic arranging of *appropriate* events to ensure the *appropriate* conditions of learning—both internal and external—are in place [4]. And, as Ertmer and Newby [2] explain, there are three main approaches to instructional design: behaviorism, cognitivism, and constructivism. The next sections will outline the key features of each approach.

7.2.1 The Behaviorist Approach

The *behaviorist* approach, otherwise known as *stimulus/response psychology*, focuses only on observable or overt behaviors and are not interested in internal processes, as they see the mind as a "black box." Thus, behaviorists study *what* people do (external actions) and do not try to determine *why* individuals engage in such behavior (internal motivators) [5].

The behaviorist approach has evolved in interesting ways over the years, and perhaps its most famous, recent proponent was the American psychology teacher B.F. Skinner (1904–1990), who gained initial fame for examining such conditioning relationships through his operant conditioning experiments involving animals. Through these experiments, Skinner found that the re-occurrence of a certain type of behavior depends upon the consequences of that behavior [5]. Put another way, if a particular behavior is followed by a reward (response), then there is a greater likelihood of that behavior re-occurring in future.

In later years, Skinner applied his research on behavior to the teaching of children. He devised the concept of "programmed instruction," which was designed to reinforce behaviors that are moving closer to the desired target behavior. Programmed instruction involves the careful sequencing of content into small steps, or chunks. Prompts are provided in the early stages of learning but are reduced as performance improves [5].

Feedback is provided on an on-going basis, rather than at the end of a lengthy session. For example, when teaching children to raise their hands when they wish to ask questions, the teacher explicitly tells the children to raise their hands each time, until they eventually learn to reflexively raise their hands, without being instructed to do so.

The principles behind Skinner's programmed instruction are still integrated into modern-day e-learning courses. Many of today's online courses, for example, are carefully structured and sequenced into modules, units, and lessons. Such structuring mimics Skinner's concept of programmed instruction, whereby content is chunked into more manageable sections. Additionally, such courses offer regular interactions throughout as well as facilities for immediate feedback. In so doing, this addresses Skinner's ideas of regular reinforcement in the early stages, which is later withdrawn as learners become more proficient.

7.2.2 The Cognitivist Approach

The cognitivist approach emerged in the 1960s as behaviorism started to become less popular. At that point in time, pedagogical theorists started to wonder about the inner processes and the reasoning *behind* overt behavior. A number of predominant theorists—including B.S. Bloom and R.M. Gagné—soon emerged to offer a different, non-behaviorist approach, to behavior and education. American educator and university examiner B. S. Bloom (1913–1999), for example, was famous for his three domains of learning and his taxonomy of educational objectives [6, 7].

Through his work, Bloom identified three major domains or types of learning—the *cognitive*, *affective*, and *psychomotor* domains. According to Bloom, cognitive skills typically refer to the kinds of intellectual skills individuals learn through formal education, such as language and mathematical skills. Affective skills, on the other hand, refer to attitudes and values. Gagné et al. [3, p. 48] define an attitude as "a persisting state that modifies the individual's choices of action." Such affective skills are taught to general practitioners, for example, who need to be able to communicate complex medical information to clients, whilst also maintaining appropriate "bedside manner." Psychomotor skills, in turn, refer to physical and motor skills and include skills such as throwing a ball, driving a car, or typing on a keyboard. Within educational contexts, it is often necessary to teach cognitive skills along with psychomotor skills, as the learner needs *knowledge* of the procedure before he/she can physically carry out that procedure. Bloom's taxonomy of educational objectives was devised as an aid teachers could use to devise appropriate assessments for their students.

For each of the three learning outcomes, Bloom and his colleagues identified a number of levels of behavior. The cognitive domain, for example, has six levels ranging from basic "knowledge" to the more advanced "evaluation." Each level is increasingly complex and mastery of previous levels is advised [8]. So, before a learner can "comprehend" the content (level 2, cognitive domain), he/she must "know" the content (level 1). Once a teacher is aware of the level of learning outcome that is desired, he/she can choose from a predefined set of outcome-illustrating verbs. For example, if a teacher wishes to examine "knowledge" (level 1, cognitive domain), he/she could ask the students to *define*, *identify*, *list*, or *recite* content. On the other hand, if a teacher wishes to examine "evaluation" skills (level 6, cognitive domain), he/she could ask the students to *conclude, critique, evaluate,* or *judge* content [6]. Bloom's work was used as a foundation for later instructional design theories, including those of Robert M. Gagné.

Robert M. Gagné, considered by many to be the founder of instructional design, initially started out as a behaviorist but later became a cognitivist. He is most famous for his conditions of learning model and the nine events of instruction associated with that model. Building on Bloom's three domains of learning, Gagné devised five types of learning outcomes, two of which (attitudes and psychomotor skills) are similar to Bloom's affective and psychomotor domains. Gagné, however, divided Bloom's cognitive domain into three sub-types: intellectual skills, verbal information, and cognitive strategies.

According to Gagné, intellectual skills are mainly comprised of rules, concepts, and discriminations and relate to "mainstream" topics such as language and mathematics. To understand a rule (e.g., milk is a good source of calcium), one must understand the concepts that make up that rule (i.e., milk, good, source, calcium). One must also have basic language skills to be able to understand the sentence construction (e.g., milk is the subject of the sentence). Likewise, to understand concepts, one must be able to understand the discriminations of that concept, where they exist (e.g., low-fat versus full-fat milk).

Verbal information relates to any label, fact, or body of knowledge that is known in sentence form (e.g., jet engines generate noise) [4]. It is important to highlight here that a learner might be able to state a verbal information fact without understanding it, which is why intellectual skills are often taught with verbal information skills. Cognitive strategies are more difficult to observe as they relate to any strategies used by learners to regulate their own learning. Such strategies might include the use of mnemonics.

Gagné's nine events of instruction were devised to facilitate teachers in the design of instruction. These nine events are

1. Gain the learner's attention
2. Inform the learner of the objective
3. Stimulate the recall of prerequisites
4. Present the stimulus material
5. Provide learning guidance
6. Elicit learner performance
7. Provide feedback on the performance
8. Assess the performance
9. Enhance retention and transfer of learning

According to this model, whilst the order of the events is not fixed, some events must naturally come before others. Feedback (event 7), for example, cannot be provided until learner performance (event 6) has been elicited.

To ensure effective learning, these nine events need to be designed and arranged to suit the audience and the type of learning outcome that is to be achieved [4]. For example, for mature learners who tend to be highly motivated and aware of the desired learning outcomes, it might not be necessary to explicitly state the objectives at the start of the lesson. Where younger learners are concerned, instructional designers might need to place *particular* emphasis on gaining (and maintaining) the learner's attention, an event that is often not required for mature learners.

7.2.3 The Constructivist Approach

The constructivist approach emerged in the 1990s, when there was a shift away from viewing learners as merely input-storage-retrieval agents (i.e., as people who receive information, store it in memory, and only recall that information when it is needed). Whilst the constructivist approach emerged from cognitivism (it still concerns itself with internal processes), it has shifted its focus to a *different* type of internal learning. The idea works as follows: Rather than acquire knowledge from external sources, the learner in a constructivist learning environment (CLE) must continuously construct his/her knowledge and understanding based on prior knowledge and experiences [9]. As a result, the teacher needs to provide a context for new learning and must relate new learning to learners' pre-existing knowledge. Only through creating such connections to previous knowledge can new information be stored for the long term—and in meaningful ways—in an individual's memory.

Additionally, the constructivist approach supports various methods of learning, including problem-based collaborative learning, library research, and fieldwork. Many virtual learning environments, such as those created by using technologies such as Sakai and Moodle, support the constructivist perspective. Facilities such as live chat rooms, discussion forums, and resource areas are provided for students and teachers alike. Using such facilities, students can become actively involved in their learning, constructing new understanding as they engage in online dialogue and problem solving with peers. Moreover, the teacher's role moves away from being an instructor to being more of an architect and facilitator of that continuous learning [10].

Additional strategies that are often employed in a CLE include

- Modeling: The teacher is expert and demonstrates to learners how and why one performs a certain task in a particular way.
- Coaching: The teacher observes the student as the student performs the task.
- Scaffolding: The teacher provides guidance and support, particularly in the early stages, but gradually removes the scaffolds as the student develops [9].

Other related theories that are closely aligned with the constructivist approach include the depth education model [11]. According to Weigel, deep learning is learning that "promotes the development of conditionalized knowledge and metacognition through communities of inquiry" [11, p. 5]. Conditionalized knowledge, in turn, is knowledge that is contextualized (i.e., learners are aware of the contexts in which it is useful). Teachers can facilitate conditionalized knowledge by asking students to undertake meaningful assignments that have relevance to their career goals or previous experiences.

Metacognition relates to the art of thinking about thinking (i.e., reflecting on why you think what you think), and such processes can be facilitated by encouraging students to maintain reflective learning blogs, for example. Communities of inquiry, also known as communities of practice, are facilitated in a learning environment when there are three types of presence—cognitive, social, and teaching presence [10]. Garrison and Anderson state that "a community of learners is an essential, core element of an educational experience when higher-order learning is the desired learning outcome" [10, p. 22].

According to this perspective, teachers need to develop skills in balancing individual and collaborative activities. CLEs such as Sakai® and Moodle® can facilitate both types of activities. They allow students to discuss topics in discussion forums while also letting them submit individual assignments directly to teachers. Garrison and Anderson [10] devised a list of categories for each of the three presences, to facilitate teachers who wish to look for evidence of deep learning. In addition to engaging in reflection and discourse (cognitive presence), students need to be able to project themselves online as real people (social presence). There is also a need for a teacher to design, organize, and facilitate the learning (teaching presence). It is the general consensus amongst teachers that no one pedagogical strategy is suitable for all learning situations. Oftentimes, it is necessary to use a combination of strategies.

In the MA program at the University of Limerick, pedagogical approaches build on the previous research as the instructors employ many of the aforementioned strategies in various combinations (see Section 7.5 for a discussion). Instructors in the program have also evaluated the impact of these pedagogical strategies on their students' learning outcomes and future careers (see, for example, References 12–15). For such practices to be successful, however, the teacher needs to be aware of the latest trends in e-learning pedagogy and needs to know how and when to adopt them. The students also need to know how and when to adopt the various strategies when they enter the workforce as e-learning developers or technical writers. If graduates are designing courses or documentation for an international audience, they will need to revise their pedagogical strategies accordingly. The next section of this chapter will discuss how instructors integrate intercultural communication pedagogies into the MA program at the University of Limerick.

7.3 Intercultural Communication Pedagogies

People in the Western world inhabit increasingly globalized societies and economies. Many corporations now organize work virtually, independent of time, space, and organizational boundaries [16]. Meanwhile, popular culture references are often the same for Irish and American teenagers. Friedman refers to the powerful force of "Americanization via globalization" as driven by old and new media, especially through online technologies and the social Web [17, p. 479].

In addition, in the past two decades, the Internet has transformed pedagogical approaches to, and indeed opportunities for, international collaboration. Social networking tools like LinkedIn, Facebook, and the many other Web 2.0 technologies now available have reduced communication barriers further. In fact, Web 2.0 technologies have expanded the range of opportunities for people from different cultures to collaborate for social, professional, and educational purposes. As Greenhow et al. [18, p. 247] observe,

Web 2.0 applications "enable hybrid learning spaces that travel across physical and cyber spaces according to principles of collaboration and participation."

Within this globalised context, it is important to ensure that future technical communication and e-learning professionals increase their awareness of communicating effectively with people from other cultures. Furthermore, technical and professional communication programs need to include content on globalization in order to prepare future graduates for globalized work environments and contexts [19], and to understand the potential consequences of corporate re-structuring [20].

The literature on international and intercultural communication pedagogy suggests many important curricular strategies can be used to accomplish such objectives. At the University of Limerick, we focus on enabling students to work individually and in teams to design training materials that are accessible to many audiences, that are internationalized, and that can be localized. This section examines three pedagogical strategies relevant for students of technical communication and e-learning:

- Collaboration
- Information design
- Rhetorical awareness

Knowledge of these three areas is crucial to help students succeed in today's globalized work environment.

7.3.1 Collaboration

Collaboration is a crucial task for technical communicators and instructional designers who must work in cross-functional teams [15, 20, 21, 22] and gather information from subject matter experts. Most corporations now manage work across geographical regions; thus workers must often operate in virtual teams [23]. Such collaboration involves relying on technology (usually lean media) to communicate with colleagues in other times zones. Despite not meeting face-to-face, virtual team members must develop trust in order to work effectively [24]. Because technical writers/professional communicators and instructional designers increasingly work in distributed, global environments [25], intercultural understanding is essential for effective collaboration [26]. Our program at the University of Limerick includes a virtual team project to enable students to collaborate virtually with students from the United States (see Section 7.6.1).

7.3.2 Information Design

Technical communicators and instructional designers are regularly responsible for designing and formatting content. One of the tenets of international information design is accessibility. In fact, the Web Content Accessibility Guidelines (WCAG 2.0) developed by the World Wide Web Consortium [27] recognize that language and culture-specific design are impediments to accessibility. Likewise, organizations like the Trace Research and Development Center acknowledge the universal benefits of accessible design [28].

Many of these guidelines speak directly to the concerns of intercultural communication teachers. In particular, Guideline 3.1 requires designers to "[m]ake text content readable and understandable." This guideline includes sub-guidelines such as:

- Specifying the human language of the page (to facilitate automated translation).
- Defining unusual words and identifying language use that is culture-specific (idioms, for example) or restrictive (such as jargon).
- Spelling out abbreviations.
- Explaining how to pronounce words, if pronunciation is central to understanding the meaning of the content.

Even guidelines that do not refer to international audiences can make online media more accessible to users from different language backgrounds. Guideline 1.1, for example, requires that images, video, audio, and other non-text content have text labels to explain the content to users who cannot perceive the component. Because it is possible to translate text labels, this guideline also benefits non-native speakers of the language. Guideline 1.2, in turn, requires designers to "provide alternatives for time-based media." In essence, this guideline serves non-native speakers by requiring captions and scripts for video and audio content.

Guideline 1.4, "[m]ake it easier for users to see and hear content including separating foreground from background," gives users control of color schemes and some layout features, including typography. This guideline serves international audiences by enabling them to adapt an interface to the design conventions of their own cultural background. Additional guidelines that benefit international audiences include those that require navigation and content to be predictable and malleable.

These guidelines enable technical and professional communicators and instructional designers to design consistently internationalized content for various media.

7.3.3 Rhetorical Awareness

In the Preface to *The Global English Style Guide* [29, p. xiii], John Kohl lists technical writers, course developers and training instructors among those professionals who need to write for translation, and for global audiences. Kohl lists several stylistic problems which inhibit translation and cause comprehension difficulties for non-native speakers. The primary problems include ambiguous phrasing, difficult or unclear sentence structures, inconsistent phrasing and presentation, and uncommon technical terms.

Style guides such as Kohl's are a useful tool for content developers, for they encourage a consistent and usable style. Another strategy to ensure that content is standardized is the adoption of a controlled language, especially in specialized fields, such as Simplified English for the aeronautical industry.

> As well as adapting language to enable translation, technical writers and instructional designers must be aware of other cultural factors that can inhibit or enhance communication. St.Amant [30, p. 47] points out that "each cultural group tends to have its own specialized knowledge related to what constitutes an acceptable and a credible presentation of information." It is possible, he argues, to analyze the communication context and use "patterns and strategies" [30, p. 61] that are appropriate for the audience.

An understanding and awareness of this linguistic and cultural diversity enables technical and professional communicators and instructional designers to create more effective internationalized content. Although we are preparing our students to work in globalized work environments, most go on to work in Ireland. In the next section, we explain the synergies between technical communication, e-learning and localization in Ireland, which led to the development of our current program in technical communication and e-learning.

7.4 The Irish Context for Technical Communication and E-Learning

Despite the homogenization of world labor markets due to globalization, the Irish labor market for knowledge workers is unique in a number of respects, and these unique traits inform our approach to teaching technical communication and e-learning at the University of Limerick. This section examines areas that differentiate Ireland from other countries and specifically examines the economic context in which corporations benefit from tax incentives. The section also considers synergies within the localization, e-learning and technical communication sectors of the Irish software industry.

7.4.1 Economic Context

Until the mid-twentieth century, Ireland was a largely agrarian society. The industrial revolution had only a limited impact on Irish economic development, and only in a small number of sectors, notably brewing and linen [31]. As a result, Ireland did not have a tradition of technical communication in support of manufacturing industries, such as was evident in heavily industrialized countries like Britain, Germany, and the United States. The Irish economy was weak for most of the twentieth century, but policies adopted in the 1960s and 1970s to attract inward investment began to bear fruit in the 1980s and 1990s. European Union (EU) subsidies enabled infrastructural improvements, while tax incentives for multinational companies, and in particular Ireland's low corporation tax rate of 12.5%, resulted in foreign direct investment (FDI).

Economic growth was especially strong in the period between 1997 and 2007 [32], and software development was one of the primary drivers of the Irish economic success story, which became known as the "Celtic Tiger" economy. In 2003, the Irish IT sector had 24,000 employees, revenues of $18 billion, and exports of $17.3 billion [33]. Multinational companies such as Microsoft and Symantec began to offshore several development functions, including technical/professional communication and most notably localization, to Ireland in the mid-1990s as Ireland was seen as a low-cost economy relative to the United States or even the United Kingdom. A report from the ACM Job Migration Task Force [34, p. 2] explains such offshoring practices in the following way:

> Over the past decade, low-wage countries such as India have developed vibrant, export-oriented software and IT service industries. Attracted by available talent, quality work, and most of all low cost, companies in high-wage countries [...] are increasingly offshoring software and service work to these low-wage countries.

Other reasons for selecting Ireland as a location for IT activity were its strong links to Europe as a member of the EU, coupled with an educated and native English-speaking workforce. As a result, Ireland became "the richest country in the European Union after Luxembourg" [17, p. 417].

Localization has become a hugely important sector for software and hardware companies, for, as Schäler [35, p. 5] explains, "[t]he overwhelming majority of publishers in the digital world now make more money from the sales of their localized products than they make from the sales of the original product." In the mid-1990s, Ireland became a world hub for localization activities [36].

While Ireland benefited from policies of off-shoring by multinational companies in the 1990s and early 2000s, the logical extension of this policy is that functions are moved to still lower-cost economies when expedient. Indeed, some technical and professional communication work is being off-shored from the United States to countries such as China and India (see, for example, References 37, 38). Hailey et al. [39, pp. 125–126] observe that "if software programming goes to a remote site with an English-speaking population, there are few valid reasons design of its entire communications package cannot follow."

It is too early to tell what impact the global economic crisis of 2008 onwards will have in the longer term on the Irish software industry and the technical/professional communication and e-learning sectors of that industry. It is noteworthy, however, that, even in this beleaguered economy, the demand for knowledge workers in Ireland is healthy [40] and content developers continue to have employment opportunities. Enterprise Ireland [41, p.2], the Irish state development agency for industry, notes that "[a]lmost all of the world's leading software companies, including IBM, Oracle, Microsoft, Google, and Facebook, have a significant presence in Ireland." The same report notes a strong indigenous software sector in Ireland, and ICT remains one of the few growth sectors in the Irish economy at present.

7.4.2 Technical Communication, E-Learning and Localization within the Irish IT Industry

Where technical/professional communication, e-learning and localization were unknown in Ireland in the 1980s, these fields are now emerging sectors of the Irish IT industry. The development of these sectors is due to the central location of IT industries, especially software development companies, in Ireland's economy from the early 1990s onwards.

Because these disciplines are new to Ireland, no workplace history of them exists in Ireland, and only anecdotal information is available on how the fields have evolved. This anecdotal evidence points to a cross-over between e-learning, localization and technical/professional communication. Most Irish technical/professional communicators, for example, now work in the software and telecommunications sectors. Some large multinational companies, including SAP and IBM, have large teams of technical/professional communicators (known as *information developers*) at plants in Cork and Galway. Additionally, e-learning companies throughout Ireland employ both technical/professional communication and instructional design specialists, many of whom are graduates of the University of Limerick's MA program.

Within Ireland's modern business context, technical/professional communicators are often involved in the localization process in a number of respects. Localization "refers to the **adaptation** of a product, application or document content to meet the language, cultural and other requirements of a specific target market (a *locale*) [42, original emphases]." The process for software localization typically involves adapting the interface, online help and manual. One of the steps in the localization process, known as internationalization, is the preparation of content that can be easily translated [43]. Technical/professional communicators, in turn, are often hired to "internationalize" text, graphics, and increasingly multimedia content for distribution to larger global audiences. Likewise, significant cross-over exists between technical/professional communication and e-learning. Both functions involve preparing clear, manageable, and accessible learning materials (often materials relating to technology use) for diverse audiences. As Carliner [44] reports, technical and professional communicators have sought to diversify into e-learning development roles in the past few years, particularly in light of the off-shoring of traditional technical communication tasks.

Some skills are common across the technical/professional communication, e-learning, and localization sectors. For example, working with graphics in structured authoring (especially XML environments), attention to culture-neutral writing and graphics, and the associated need for plain English are essential to success in all three areas. In an acknowledgement of the intersections between these fields, the International Professional Communication Conference (IPCC) committee invited Reinhard Schäler, Director of the Localization Research Centre at the University of Limerick, to present a keynote address at the 2005 IEEE IPCC. In addition, localization regularly features on the lists of topics in calls for papers for technical communication and e-learning conferences. The University of Limerick's MA program thus acknowledges these synergies and is framed within the economic and occupational context described above.

7.5 The Configuration of our Program

The MA in Technical Communication and E-Learning at the University of Limerick is the result of the merger of two graduate programs: the Graduate Diploma/MA in

Technical Communication and the MA in E-Learning Design and Development. This section outlines the program structure, content and assignments. We outline the key attributes graduates gain, as well as more specific learning outcomes of individual courses.

7.5.1 The Structure of our MA Program

The University of Limerick's MA program focuses on content development for e-learning and technical/professional communication. The program's purpose is to equip its graduates with the skills needed to design and develop on- and off-line content, with a particular emphasis on technology-enhanced content solutions. On successful completion of the program, graduates are able to

- Design and deliver instructional materials
- Evaluate instructional tools and programs
- Use multimedia applications to create content
- Deploy and use Web 2.0 technologies to collaborate with peers
- Manage complex writing and design projects
- Communicate effectively in online and face-to-face environments
- Write clear, correct, precise content

This breadth of program outcomes ensures that our students are able to work in a variety of related fields on completion of the program. Whilst the majority of our students come from non-engineering backgrounds, graduates from the program typically find work in technical roles as instructional designers, technical writers, and e-learning course developers. As discussed in Section 7.4.1, many graduates find employment in large software development companies that develop their own e-learning courses and documentation, such as SAP® and IBM.®

The majority of students enroll in the program on a full-time basis, completing nine courses in a 12-month period; part-time students take the program over a two-year period of time. The full- and part-time program structures are outlined in Tables 7.1 and 7.2.

In the fall and spring semesters, all courses are taught on-campus. Students undertake continuous assessments, and there are no end-of-semester examinations. There are no classes in the Summer semester, however. Instead, during the summer months, each student must take either the dissertation or the development route. Students taking the dissertation route must write a dissertation on a topic that is relevant to the program (these summer dissertations are typically in the range of 15,000 words and are therefore smaller in scope than a PhD thesis or a 2-year Master's thesis). Students taking the development route must design and develop a technical or professional communication product, for example, a substantial web site, manual, or e-learning course, they must conduct a usability study on the effectiveness of the product, and they must write a short report describing the development and usability evaluation stages. In the next section of

TABLE 7.1. Example of the Full-Time Program Structure for the MA in Technical and Communication and E-Learning at Limerick University

Fall Semester	Spring Semester	Summer Semester
Principles of Professional and Technical Communication and Information Design	Workplace Issues in Technical and Professional Communication	Dissertation/Project
Instructional Design	E-Learning Theories and Practices	
Theory of Technical Communication	Interactive Courseware Workshop	
Research Methodology in Applied Language Studies	Learning and Collaboration Technologies	

TABLE 7.2. Example of the Part-Time Program Structure for the MA in Technical and Communication and E-Learning

Fall Semester	Spring Semester	Summer Semester
Year One:		
Principles of Professional and Technical Communication and Information Design	Workplace Issues in Technical and Professional Communication	
Instructional Design	Interactive Courseware Workshop	
Year Two:		
Theory of Technical Communication	E-Learning Theories and Practices	Dissertation/Project
Research Methodology in Applied Language Studies	Learning and Collaboration Technologies	

this chapter, we will briefly outline the content of the taught courses that have intercultural communication, internationalization, or localization components. Courses will be discussed in the order in which they are undertaken by students.

7.5.2 The Courses in our MA Program

In the course *Principles of Professional and Technical Communication and Information Design*, students learn about color, typography, how to write in clear, concise, correct English (our focus is British English) and how to use appropriate styles for a range of genres and diverse audiences. Students are required to write summaries, rewrite passages, critique and redesign graphics, design brochures, and develop instructional manuals. When these students enter the workforce, they will be required to convert complex technical documents into more readable documentation, oftentimes for international audiences. The afore-mentioned assignments are therefore designed to prepare students for working in such a dynamic environment. (A more detailed discussion of some of the writing assignments can be found in Section 7.6.3.)

In the course *Instructional Design*, students learn about pedagogical theories and different approaches to instructional design, including the systematic design of instruction model [45]. When teaching this latter model, the instructors cover audience analysis in detail. The instructors also teach students about localization and internationalization considerations that must be considered when designing and developing instructional materials that may subsequently be used by international audiences. For example, the class teaches students about color, icons, and the importance of using Global English in e-learning courses and web sites that can be accessed by a global audience. In this course, students are asked to write essays demonstrating their in-depth knowledge of the theories and how they might be applied in e-learning course development.

In the course *Theory of Technical Communication*, students learn about the literature that underpins technical communication practices. They also learn about information design theory and how different types of users interact with various products and content. The major assignment for the course requires students to carry out a research project on a topic of relevance to technical communication, and the majority of students carry out usability studies on products or documentation. As part of this course, students are also required to participate in online discussions with students enrolled in the online Graduate Certificate in Technical Writing program. Because many of our distance students never meet their classmates face-to-face, these online discussions give our on-campus and distance students opportunities to collaborate with, and learn from, students who have different cultural and experiential backgrounds.

In the course *Workplace Issues in Technical and Professional Communication*, students learn about ethical and legal considerations that must be kept in mind by professional communicators. Students also learn about interpersonal, intercultural, and nonverbal communication. The individuals enrolled in the class are required to interview a professional working in a relevant field and to report on the findings orally and in an electronic presentation. Finally, the students participate in online discussions with distance students, and such interactions provide opportunities to improve cross-cultural communication skills.

In the course *Interactive Courseware Workshop*, students learn practical design and development skills, such as how to develop e-learning courses, web sites, and animations, using industry-standard tools (e.g., Adobe Dreamweaver and Flash). In all assignments, students are required to demonstrate their knowledge and understanding of accessibility issues and international audiences. For example, students are expected to use culturally-appropriate color schemes, icons, and Global English in their assignment submissions, whenever possible. They are also required to submit written reports, which describe how and where they applied the relevant theories to the design and development stages. (A more detailed discussion of some of these assignments is presented in Section 7.6.2.)

In the course *E-Learning Theories and Practices*, students learn how early distance education has evolved into e-learning, and more recently, mobile learning (m-learning). The instructor and the students discuss the issues facing teachers who are charged with teaching students who are located in geographically-dispersed locations. The instructor also teaches students about ethical and access issues faced by distance learning students. For example, *potential* distance students might not have access to the latest technologies and/or might not have the expertise to use the latest e-learning technologies. As a result, they might be prevented from participating in distance education.

The course introduces students to issues that must be considered when setting up a learner-support-system for distance learners. Individuals also learn how to e-moderate online discussions, with a particular emphasis on netiquette and using appropriate communication strategies when dealing with international students. Knowledge of the various issues and challenges faced by teachers and distance students is crucial, particularly as teachers can no longer assume that all students possess the same cultural backgrounds and language capabilities. After all, it is now critical that online teachers are aware of the issues associated with teaching international audiences. For this reason, students enrolled in this course are asked to write essays demonstrating their in-depth knowledge of the theories and how they might be applied in e-learning course development. They are also required to participate in online discussions, where they discuss how they might deal with such issues.

In the course *Learning and Collaboration Technologies*, students experiment with a range of new technologies and evaluate the advantages and disadvantages of using these technologies for communication and learning. Typical assignments include a reflective learning blog, an educational or entertaining podcast, and a virtual team assignment. It is likely that our students will have to use a variety of technologies to collaborate in global virtual teams, when they enter the workforce.

7.6 The Assignments in the MA Program

This section will describe three types of assignments we issue in the University of Limerick's MA program, and each assignment has an emphasis on learning pedagogies and/or communication for international audiences. For the different assignment types, we outline the objectives of the assignment as well as provide typical evaluation criteria. In so doing, we also emphasize the relevance of the assignment type to the workforce/workplace contexts. These three types of assignments (virtual team, design and development, and writing assignments) were selected for examination here because they represent a cross-section of assignments undertaken by students, requiring them to demonstrate different global communication skills.

7.6.1 Virtual Team Assignments

Over a period of 5–7 weeks in the spring semester, students at the University of Limerick (UL) and University of Central Florida (UCF) are required to use a range of technologies to communicate and collaborate on a particular assignment. In the spring of 2009, for example, students were required to write a joint-authored research report on different collaborative technologies. Each team member was responsible for selecting a different technology and proposing it to the participating instructors for inclusion. Each team, comprising both Irish and US students, then had to use the proposed technologies to engage in collaborative work throughout the project.

This assignment required students to (1) identify interesting and relevant technology, (2) to use these technologies for collaboration, and (3) to report on the strengths and weaknesses of each technology for collaborative work. Each student was also required to write about his/her technology in the research report.

Instructors at UL and UCF awarded points based on the following evaluation criteria:

- *The content of the report.* The coverage of each tool had to be consistent and students had to analyze the advantages of each tool in-depth. Students also had to refer to relevant journal articles, white papers, and textbooks, as well as their own experiences using the tools.
- *The presentation of the information.* Points were awarded for well-organized content and adherence to the report template provided. Even though each student was required to write about his/her tool, the report had to have a unified voice and the writing style needed to be consistent.
- *Their professionalism and team participation.* Each team was expected to have at least two meetings per week and we facilitated one of these meetings during class time. Students were free to use whichever tools they wished for collaboration, but they had to ensure that their instructors had access to their collaborative notes and meeting minutes.

In addition to a team grade for collaboration, students were given individual grades for their participation in these activities. The UL instructors assigned Irish students a grade based on reflective learning blogs, which each student was required to maintain as part of another assignment in the Learning and Collaboration Technologies course (see Section 7.5.2). The US instructor graded US students based on their individual wrap-up reports.

This assignment requires students to consider the impact of cultural communication decisions, such as writing style and information layout, and to develop their online communication skills. As discussed in Section 7.3.1, the majority of these students will be required to participate in virtual teams when they enter the workforce, and this assignment gives them valuable experience of the challenges typically faced when working with teammates who are in different geographic locations.

7.6.2 Design and Development Assignments

Students in the University of Limerick's MA program are also required to undertake a number of design and development assignments. One such assignment, the web site assignment, requires students to work in teams to design and develop a web site for a real-world client. This assignment has a number of components and objectives. First, it assesses each student's ability to write a collaborative proposal, whereby students have to agree on, and outline, the writing style, interface design, and interactive features that will be employed in the final web site. Second, the assignment assesses each student's technical proficiency in using industry-standard tools (in this case, Adobe Dreamweaver) to collaboratively develop a web site. Finally, it assesses each student's ability to collaborate with peers on a real-world project.

In terms of evaluation criteria for the group proposal, grades are awarded for writing style, clarity of proposal ideas, and overall presentation of the proposal document. Even though each team member must contribute to the proposal, the proposal must be presented in a unified voice. The final web site is then graded based on

- Suitability of the content for the various stakeholders, some of whom are likely to be from other countries
- Presentation of text layout
- Appropriate use of color and graphics
- Appropriate use of interactive features
- Demonstrated proficiency in using Adobe Dreamweaver

As can be seen from the objectives outlined earlier in this section, this assignment is directly transferrable to the workforce. When our graduates gain employment as instructional designers or technical writers, they will most likely be required to work in teams to write documentation and/or to develop web sites and e-learning courses. These workplace products will be all the more useful if students apply the theories they studied in the program. For that reason, we require students to *justify* their design and development choices, in the initial team proposal.

7.6.3 Writing Assignments

Writing pedagogy is a cornerstone of our program, because technical communicators and instructional designers must be adept at writing clearly and producing content to required specifications, often within short timeframes. Our pedagogical approaches recognize the influence of culture, and the importance of cultural awareness when making decisions about, for example, word choices and sentence structure, idiomatic phrasing, grammatical constructions, and terminology choices.

Students complete a variety of writing-related assignments in the MA program. In the first semester, they write summaries, critique documents, write essays, and rewrite poorly written texts, including manuals. In the second semester, they design a reflective learning blog, a written report that outlines the findings of an interview with a professional writer, and a number of essays. Beginning with smaller writing assignments in the first semester and gradually building up to more challenging, reflective projects in the second half of the program, enables students to develop their skills and to become more confident writers. In addition to honing their writing skills, these assignments help to attune students to the requirements of writing for different audiences. As well as writing-specific assignments, during both semesters, students must design well-written and audience-appropriate content for their development projects such as web sites and e-learning courses, as discussed in Section 7.6.2.

7.7 Connecting Student Work to Different Contexts

The following sub-sections focus on three writing assignments completed by our students. We examine them in terms of their workplace applicability, learning outcomes, assessment criteria, and student feedback. These assignments are rewriting a poorly written text, critiquing and redesigning an instruction manual, and writing a reflective learning blog.

7.7.1 Rewriting a Poorly Written Text

In the Fall semester, as part of the course *Principles of Professional and Technical Communication* (see Section 7.5.2), one of the first assignments requires students to rewrite a poorly-written text. The text is about 500 words long, and typically contains several stylistic problems that make it difficult for both non-native and native speakers to understand. (Such stylistic problems include overuse of passive constructions, imprecise terminology, very long sentences with multiple clauses, negatively worded phrases, and so on.)

This assignment is applicable to the workplace because a large part of the technical writer's and instructional designer's role involves converting technical specifications and procedures into clear, lucid language that is suitable for the target audience. In many cases, the content will have to be translated into other languages and be localized for other cultures. Therefore, future content developers must recognize constructions that regularly cause problems for translators, including passive constructions, clichés, idioms, negative phrasing, unconventional word usage, and unnecessary phrasing (see, for example, References 29 and 46).

The objectives of this assignment are to enable students to analyze the target audience and to write clear, correct, concise, lucid, English suitable for the target audience. The instructors assess the work in terms of how explicitly students have converted the text into a format that is usable and translatable. Feedback from graduates indicates that this assignment is valuable in preparing them to apply for technical writing and instructional design positions where applicants are often required to rewrite a text to determine their suitability for the role.

7.7.2 Critiquing and Redesigning an Instruction Manual

In the same course, the final assignment requires students to critique and redesign an instruction manual. Students may choose from four possible manuals, each with problems in the areas of content, layout, organization, writing style, and graphics. Students must address each problem area within the critique, using examples to illustrate their points. They must also refer to relevant readings to show that their critique has a theoretical foundation.

This assignment is relevant to the future work of technical writers and instructional designers, who in most roles in which they function must write procedural information for different audiences as part of their work. The objectives of the assignment are to enable students

- To analyze an existing manual
- To critique aspects of its layout, design and content
- To rewrite it in a style and format suitable for the target audience

For this assignment, students must deliver a hard copy of the manual; thus, a further objective is to design a high-quality deliverable.

This assignment assesses many intercultural communication criteria. For example, students must design clear, lucid English, in a well-organized document that includes navigation devices such as a Table of Contents, appropriate graphics, and a clear, usable format. Feedback from students indicates that this assignment is challenging because it tests their writing, information design, and organization skills.

7.7.3 Writing a Reflective Learning Blog

In the spring semester, as part of the course *Learning and Collaboration Technologies* (see Section 7.5.2), students design a reflective learning blog. They must design at least two blog entries per week, on topics relevant to their program of study. Typical topics include ethical issues; the online learning environment; technologies for e-learning; culture and learning; and information design.

This assignment has workplace applicability by providing an opportunity for students to contribute to the discourse community of their subject area. The act of blogging can represent a contribution to the profession and to the community of practice [47]. Research on blogging shows that this communication medium is becoming more focused, and many discipline-specific blogs are emerging [48]. Moreover, a student's blog can also become a component of his/her portfolio, and can distinguish applicants from their peers in a job search. Considering that competition for appropriate jobs is likely to increase during a recession, this advantage is significant for our graduates.

The objectives of this assignment are to

- Enable students to exploit Web 2.0 technologies
- Write for a potentially global audience
- Reflect on topics they have learned about on the program
- Reflect on the real-world application of tools and techniques
- Practice writing clear, concise, analytical prose

Assessment of the blog is, in turn, based on the number of blog entries, the level of reflection, the range of topics covered, the accessibility of the writing style, and the treatment of divergent and emerging issues in the discipline. Feedback from students indicates the tensions implicit in this genre: the pleasure of writing and reflecting on professional topics combined with the challenges of writing for a global audience.

7.8 Conclusion

Technical communicators now work in a variety of content development fields, including instructional design for e-learning, and localization. This chapter contributes to the literature on intercultural communication pedagogy, by addressing assessment, and by instancing the synergies between instructional design, technical communication, and localization. The assignments our students complete are designed to enable them to function effectively as content developers, of e-learning and technical communication products, for global audiences.

The MA in Technical Communication and E-Learning at the University of Limerick has evolved in response to the changing industrial and technological landscape. Although our graduates are uniquely qualified to work in information industries in Ireland, they are likely to participate in global teams in the workplace. The range of assignments discussed in the chapter incorporates aspects of collaboration, online development, and writing. Feedback from students and graduates indicates that these assignments, though challenging, prepare students for consonant workplace activities.

References

1. H. Yu, "Intercultural competence in technical communication: A working definition and review of assessment methods," *Technical Communication Quarterly*, vol. 21, no. 2, pp. 168–186, January, 2012.
2. P. A. Ertmer and T. J. Newby, "Behaviorism, cognitivism, constructivism: Comparing critical features from an instructional design perspective," *Performance Improvement Quarterly*, vol. 6, no. 4, pp. 50–72, December, 1993.
3. R. M. Gagné, L. J. Briggs, and W. W. Wager, *Principles of Instructional Design*, 4th ed. Belmont, CA: Wadsworth/Thomson Learning, 1992.
4. R. M. Gagné, W. W. Wager, K. C. Golas, and J. M. Keller, *Principles of Instructional Design*, 5th ed. Belmont, CA: Wadsworth, 2004.
5. B. F. Skinner Foundation, "A brief biography of B.F. Skinner," 2012. Available at: http://www.bfskinner.org/bfskinner/AboutSkinner.html (accessed November 1, 2013).
6. B. S. Bloom, Ed., *Taxonomy of Educational Objectives. Book 1: Cognitive Domain*. New York: Longman, 1956.
7. D. R. Krathwohl, B. S. Bloom, and B. B. Masia, *Taxonomy of Educational Objectives: The Classification of Educational Goals. Handbook 11: Affective Domain*, 2nd ed. New York: Longman Publishing Group, 1999.
8. E. W. Eisner, "Benjamim Bloom," *Prospects: The Quarterly Review of Comparative Education*, vol. 30 no. 3, pp. 387–395, September, 2000.
9. D. H. Jonassen, "Designing constructivist learning environments," in *Instructional-Design Theories and Models: A New Paradigm of Instructional Theory*, vol. 2, C. M. Reigeluth, Ed. New Jersey: Lawrence Erlbaum Associates, Inc., Mahwah, NJ, pp. 215–240, 1999.

10. R. Garrison and T. Anderson, *E-Learning in the 21st Century*. London: RoutledgeFalmer, 2003.
11. Van B. Weigel, *Deep Learning for a Digital Age*. San Francisco, CA: Chichester: Jossey-Bass, 2002.
12. D. M. Slattery, "Using information and communication technologies (ICTs) to support deep learning in a third-level on-campus course: A case study of the taught Master of Arts in E-Learning Design and Development course at the University of Limerick," in *Proceedings of the IEEE International Professional Communication Conference*, Saratoga Springs, New York, 2006, pp. 170–182.
13. D. Slattery, M. Flammia, and Y. Cleary, "Preparing technical communication students for their role in the information economy: Client-based virtual team collaboration between Irish and US students," in *Proceedings of the IEEE International Professional Communication Conference*, Montréal, Canada, 2008, pp. 1–10.
14. Y. Cleary and D. Slattery, "Virtual teams in higher education: Challenges and rewards for teachers and students," presented at the AISHE International Conference Series—Valuing Complexity: Celebrating Diverse Approaches to Teaching and Learning in Higher Education, NUI Maynooth, Ireland, 2009.
15. M. Flammia, Y. Cleary, and D. Slattery, "Leadership roles, socioemotional communication strategies, and technology use of Irish and US students in virtual teams," *IEEE Transactions on Professional Communication,* vol. 53, pp. 89–101, 2010.
16. D. Robey, H. M. Khoo, and C. Powers, "Situated learning in cross-functional virtual teams," *Technical Communication*, vol. 47, no. 1, pp. 51–66, 2000.
17. T. L. Friedman, *The World is Flat: A Brief History of the Twenty-First Century*. New York: Picador, 2007.
18. C. Greenhow, B. Robelia, and J. E. Hughes, "Learning, teaching, and scholarship in a digital age: Web 2.0 and classroom research: What path should we take now?" *Educational Researcher*, vol. 38, no. 4, pp. 246–259, May, 2009.
19. B. Thatcher, "Intercultural rhetoric, technology, and writing in Mexican maquilas," *Technical Communication Quarterly*, vol. 15, no. 3, pp. 383–405, Summer, 2006.
20. J. Jablonski, "Seeing technical communication from a career perspective: The implications of career theory for technical communication theory, practice, and curriculum design," *Journal of Business and Technical Communication*, vol. 19, no. 1, pp. 5–41, January, 2005.
21. P. E. Brewer, "Miscommunication in international virtual workplaces: A report on a multicase study," *IEEE Transactions on Professional Communication,* vol. 53, pp. 329–345, December, 2010.
22. R. Spilka, *Digital Literacy for Technical Communication: 21st Century Theory and Practice*. New York: Routledge, 2010.
23. L. Martins, L. L. Gilson, and M. T. Maynard, "Virtual teams: What do we know and where do we go from here?" *Journal of Management*, vol. 30, no. 6, pp. 805–835, December, 2004.
24. N. W. Coppola, S. R. Hiltz, and N. G. Rotter, "Building trust in virtual teams," *IEEE Transactions on Professional Communication,* vol. 47, pp. 95–104, June, 2004.
25. C. Spinuzzi, "Guest editor's introduction: Technical communication in the age of distributed work," *Technical Communication Quarterly*, vol. 16, no. 3, pp. 265–277, 2007.
26. D. Andrews and D. Starke-Meyerring, "Making connections: An intercultural virtual team project in professional communication," in *Proceedings of the IEEE International Professional Communication Conference*, Limerick, Ireland, 2005, pp. 26–31.

REFERENCES

27. W3C. (2008, Dec. 11). *Web content accessibility guidelines* Available at: http://www.w3.org/TR/WCAG/ (accessed October 30, 2013).
28. Trace Research and Development Centre. *Research to Make Everyday Ttechnologies Accessible and Usable...* March, 2010. Available at: http://trace.wisc.edu/ (accessed July 30, 2015).
29. J. R. Kohl, *The Global English Style Guide: Writing Clear, Translatable Documentation for a Global Market.* Cary, NC: SAS Institute, 2008.
30. K. St.Amant, "Globalizing rhetoric: Using rhetorical concepts to identify and analyze cultural expectations (specialized knowledge) related to genres," *Hermes – Journal of Language and Communication Studies*, vol. 37, pp. 47–66, 2006.
31. L. Mjøset, *"The Irish economy in a comparative institutional perspective,"* National Economic and Social Council, Rep. 93, 1992. Available at: http://www.nesc.ie/en/publications/publications/nesc-reports/the-irish-economy-in-a-comparative-institutional-perspective/ (accessed July 30, 2015).
32. K. P. V. O'Sullivan and T. Kennedy, "What caused the Irish banking crisis?" *Journal of Financial Regulation and Compliance*, vol. 18, no. 3, pp. 224–242, 2010.
33. A. Sands. "The Irish software industry," in *From Underdogs to Tigers: The Rise and Growth of the Software Industry in Brazil, China, India, Ireland, and Israel*, A. Arora and A. Gambardella, Eds. Oxford: Oxford University Press, pp. 41–71, 2006.
34. W. Aspray, F. Mayadas, and M. Y. Vardi, Eds, "Globalization and offshoring of software: Report of the ACM Job Migration Task Force," ACM, New York, NY, Rep. 0001-0782/06/0200, 2006. Available at: http://www.acm.org/globalizationreport/pdf/shortv.pdf / (accessed August 1, 2015).
35. R. Schäler, *"The cultural dimension in software localisation,"* Localisation Reader 2003–2004, 5–9, 2003. Available at: http://www.localisation.ie/resources/reader/2003/-00-11%20LR-S.pdf#page=7 (accessed August 1, 2015).
36. J. Kirkman, "From chore to profession: How technical communication in the United Kingdom has changed over the past twenty-five years," *Journal of Technical Writing and Communication*, vol. 26, no. 2, pp. 147–154, 1996.
37. A. Thayer, *"Offshoring, outsourcing, and the future of technical communication,"* in Proceedings of IEEE International Professional Communication. Conference, Limerick, Ireland, 2005, pp. 567–577.
38. G.F. Hayhoe, "The globalization of our profession," *Technical Communication*, vol. 53, no. 1, pp. 9–10, 2006.
39. D. Hailey, M. Cox, and E. Loader, "Relationship between innovation and professional communication in the 'creative' economy," *Journal of Technical Writing and Communication*, vol. 40, no. 2, pp. 125–141, 2010.
40. J. Behan, N. Condon, A. M. Hogan, J. McGrath, J. McNaboe, I. Milićević, and C. Shally, "National skills bulletin 2011: A study by the Skills and Labour Market Research Unit (SLMRU) in FAS for the Expert Group on Future Skills Needs," Forfas, Dublin, Rep. 7, 2011. Available at: http://www.skillsireland.ie/media/EGFSN110706-National_Skills_Bulletin_2011.pdf (accessed August 1, 2015).
41. Enterprise Ireland, "Best connected: Software from Ireland: A strategy for development of the indigenous software industry 2009–2013," Enterprise Ireland, Dublin. Available at: http://www.enterprise-ireland.com/en/Publications/Reports-Published-Strategies/Enterprise-Ireland-Software-Strategy-2009-2013.pdf (accessed August 1, 2015).
42. W3C. *Localization vs. Internationalization*, September 27, 2010. Available at: http://www.w3.org/International/questions/qa-i18n.en.php (accessed August 1, 2015).

43. B. Esselink, "The evolution of localization," in *Translation Technology and its Teaching*, A. Pym, A. Perekrestenko, and B. Starink, Eds. Tarragona: Intercultural Studies Group, 2006, pp. 21–30.
44. S. Carliner, "Computers and technical communication in the 21st century," in *Digital Literacy for Technical Communication: 21st Century Theory and Practice*, R. Spilka, Ed. New York: Routledge, 2010, pp. 21–50.
45. W. Dick, L. Carey, and J. O. Carey. *The Systematic Design of Instruction*, 6th ed. London: Pearson, 2004.
46. E. H. Weiss, *The Elements of International English Style: A Guide to Writing Correspondence, Reports, Technical Documents, and Internet Pages for a Global Audience*. Armonk, NY: M.E. Sharpe, 2005.
47. A. Lenhart and S. Fox, "Bloggers: A portrait of the Internet's new storytellers," Pew Internet and American Life Project, Washington DC, 2006. Available at: http://www.pewinternet.org/~/media//Files/Reports/2006/PIP%20Bloggers%20Report%20July%2019%202006.pdf.pdf (accessed August 1, 2015).
48. E. Davidson and E. Vaast, "Tech talk: An investigation of blogging in technology innovation discourse," *IEEE Transactions on Professional Communication*, vol. 52, pp. 40–60, March 2009.

8

Teaching and Training with a Flexible Module for Global Virtual Teams

Pam Estes Brewer

Mercer University

Few organizations provide structured education and training in international virtual team work. In this chapter, the author describes an experience-based, global virtual team project that she has been using in junior- and senior-level professional communication classes. From this experience, she has developed a flexible module that can be used by teachers and trainers who wish to better prepare people to work effectively in global virtual teams. With the module, virtual team projects can be constructed for most classroom and training situations where they contribute to moving students along the continuum from global/online awareness to global/online competency.

8.1 Introduction

The skills needed to work successfully in virtual teams that support a great deal of global work flow are not intuitive for many people; they are gained through experience. Few organizations provide structured education and training in global virtual teamwork, and consulting organizations are beginning to fill that gap. The hiring of these consultants indicates that managers are starting to recognize the need for such education. In this chapter, I describe a global virtual team project I have been using in my junior- and

---

Teaching and Training for Global Engineering: Perspectives on Culture and Professional Communication Practices,
First Edition. Edited by Kirk St.Amant and Madelyn Flammia.
© 2016 The Institute of Electrical and Electronics Engineers, Inc. Published 2016 by John Wiley & Sons, Inc.

senior-level professional communication classes, presenting some of the results that I and my students have noted in relation to these projects.

> The skills needed to work successfully in virtual teams are not intuitive for many people; they are gained through experience.

From this experience, I have developed, and present here, a flexible module that can be used by teachers and trainers who wish to prepare people to work effectively in global virtual teams. This module is comprised of 11 critical tasks and has been applied in four cases. Each case, in turn, is described via project structure, goals, concrete experience, and implications. The module is flexible in that it can be used in classes of varying disciplines, topics, levels, sizes, etc. It is also possible, as is demonstrated in the cases here, to use the module to pair two significantly different classes. The only consistent constraint to such pairings is that the classes must be able to communicate in a single language. The virtual team projects constructed using the module presented here can be incorporated into most classroom and training situations where they contribute to moving students along the continuum from global/online awareness to global/online competency.

8.2 The Origins of the Approach Presented in This Chapter

From a strictly educational perspective, projects that involve the pairing and collaboration of students in different cultures support the internationalization efforts that are sweeping many university campuses. The flexible project presented in this chapter can

- Promote internationalization of curricula
- Increase international research and faculty development
- Expand international campus and community outreach activities
- Enhance campus and community communication

I began to develop this flexible module in cooperation with a Culture and Language across the Curriculum (CLAC) grant.

> Projects that involve the pairing and collaboration of students from different cultures support the internationalization efforts that are sweeping many university campuses.

As these projects developed, I recognized the powerful teaching tool that they are; I can educate students in a professional communication genre and at the same time help them to improve their online and global literacy—a rich combination of theory and practice. Use of the global virtual team module for teaching requires only the technology that is commonly found in most university computer labs and on the Internet. In addition, the module is consistent with experiential learning advantages wherein students make connections, are faced with compelling situations, and actively participate in learning [1].

I have used this global virtual team project effectively in my face-to-face (f2f), hybrid, and online classes. The project allows my students to practice in real-world, international situations such as those engineers and professional communicators might encounter. In the following sections, I review the basic theory of experiential learning, describe the specific projects I have launched based on this flexible module, share student and instructor observations, and describe a flexible module from which other teachers and trainers may construct such projects.

8.3 International Virtual Communication and Experiential Learning

Engineering projects are more often globally coordinated today than they were years ago, and they rely heavily on information and communication technologies (ICTs) enabling engineers to work across boundaries of space and time [2,3]. However, the pace of education and training that prepares professionals to work in global virtual environments has not kept up with the practice. In her study of training practices in a business process outsourcing firm in India, Raju [4] found little existence of and consistency in the training provided to those who work in global virtual team environments. Brewer [5] similarly found that while employees desired a range of education and training, most had received little to no structured preparation for working successfully in such environments. Furthermore, those individuals who collaborate in global virtual teams are faced with many challenges, some of which are common to all teamwork and some of which are particular to online collaboration (e.g., References 6, 7). Olaniran and Edgell [7] identify some of these challenges as language, culture, trust, and technology; they also acknowledge these challenges as "complex and intertwined" [7].

Educators are increasingly turning their attention to the deficit of consistent instruction on intercultural online communication and are examining ways to improve the preparedness of a future workforce in this area [3], [7–12]. While significant bodies of research on both intercultural and virtual communication do exist, a growing body of research is also emerging at the intersection of the two: the area of intercultural virtual communication.

It is at this intersection that educators are attempting to prepare the new workforce. For example, Zemliansky [8] has used global virtual team projects within single semester classes, while Herrington has used the global classroom project [9]. Rutkowski and colleagues [3] have used the HKNet (Hong Kong Netherlands) project to bring engineering students together in global virtual teams to build web sites that are then integrated into one large project. This project spans 10 years and a thousand students. Rutkowski et al. point out that while such training could be addressed in the workplace, a more effective approach is to begin earlier with university education. Zemliansky [8], in turn, identifies the need for education and training in both cross-cultural communication and virtual environments. Additionally, scholars such as Zemliansky, Herrington, and Olaniran (as well as myself) are examining methods of instruction and are often noting the power of experiential learning to support teaching with global virtual teams. Other scholars [3], [13], while not explicitly referring to their instructional methods as "experiential," describe methods consistent with experiential characteristics (i.e., grounded in experience and including resolution of conflict, adaptation, transactions between people and the world, and creation of knowledge).

8.3.1 Experiential Education Theory

Experiential learning can broadly be defined as learning by doing and reflecting on doing. It is based in a long history of the study of learning and development by such scholars as Lewin, Dewey, Piaget, and others [14–16]. David A. Kolb sought to synthesize experiential learning theory into a more usable whole and identified the common principles of this type of learning as summarized in the following list:

1. Learning is best conceived as a process, not in terms of outcomes
2. Learning is a continuous process grounded in experience
3. The process of learning requires a resolution of conflicts between dialectically opposed modes of adaptation to the world
4. Learning is an holistic process of adaptation to the world
5. Learning involves transactions between the person and the environment
6. Learning is the process of creating knowledge [17; ordering of this list provided by the author]

These characteristics have become increasingly important in the decades since they were articulated. Because the world changes quickly and continuously, experiential learning (which emphasizes continuous process) is particularly suited to preparing people for continuous change. Employers are calling for improved critical thinking and problem solving skills from those they hire and recognize that *real* experiences in education contribute to gaining such skills. Those engaged in engineering and the business of engineering today must be able to learn as a continuous process. Figure 8.1 illustrates Lewin's experiential learning model; while more complex models of experiential learning exist such as those by Dewey and Piaget, this simpler diagram captures the essential parts of the process.

Lewin's model captures well the common characteristics of experiential learning with its core based on participation in concrete experience, observation, reflection, adaptation, and application. The process of moving through this cycle is recursive, messy, and real—as is learning from experience in a world-of-work setting. Students will likely find themselves learning multiple lessons and adapting and testing many theories. They may move smoothly through some parts of the experience while failing and returning to other parts. In the case of the global virtual teams discussed in this chapter, students work through this cycle as a team. Kolb cites one of the greatest strengths of the experiential model when he writes, "When human beings share an experience, they can share it fully, concretely, *and* abstractly" [17]. In essence, he is noting that people often learn best together during a common experience. During the experience, they have the opportunity to share both explicit and tacit information that is assimilated into knowledge.

The experiential learning model (which can be effectively applied to both education and training) is further supported by the report of the Joint Task Force on Student

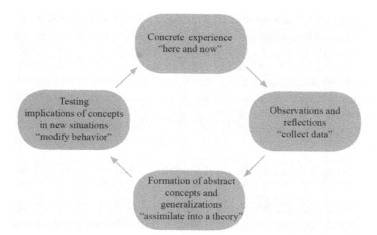

FIGURE 8.1. Lewinian experiential learning model. The Lewinian Model captures the essential parts of the experiential learning process [17]. (Description in quotation marks added by this author.)

Learning. This task force was created by the American Association of Higher Education, the American College Personnel Association, and the National Association of the Student Personnel Administrators for the purposes of reviewing a large number of student learning initiatives on various campuses. The Task Force's report emphasizes 10 principles of learning as noted in the following list:

1. Learning is fundamentally about *making and maintaining connections* ...
2. Learning is enhanced by *taking place in the context of a compelling situation* ...
3. Learning is an *active search for meaning* by the learner ...
4. Learning is *developmental*, a cumulative process *involving* the *whole person* ...
5. Learning is done by individuals who are intrinsically *tied to others as social beings* ...
6. Learning is *strongly affected by the educational climate* in which it takes place ...
7. Learning requires *frequent feedback* if it is to be sustained, *practice* if it is to be nourished, and *opportunities to use* what has been learned ...
8. Much learning *takes place informally and incidentally* ...
9. Learning is *grounded in particular contexts and individual experiences* ...
10. Learning involves the *ability of individuals to monitor their own learning* ... [1] [Ordering of this list provided by the author. Italics are preserved from the original.]

As we consider the principles of learning described in the two previous lists (the principles in the first list from people who are often identified as seminal thinkers in learning and in the second list from educational leaders 20 years later), it is striking that the core learning principles in both lists are very similar and exemplary of experiential learning. For example, the second principle of experiential learning in the first list is common to the theories of Lewin, Dewey, and Piaget. This principle emphasizes the importance of process and experience. If we then consider the principles of effective student learning in the second list, principles 1, 2, 3, 4, 7, 8, and 9 all emphasize learning through process and experience. Thus, it can be argued that both seminal theorists in education as well as expert members of the Joint Task Force on Student Learning recognize an experiential approach as an effective approach to learning.

8.3.2 Experiential Learning as it Complements Virtual Team Learning

Let us now turn to the work being done in global virtual education. Zemliansky [8] and Herrington [9] explain that experiential forms of instruction are ideally suited to teaching and learning how to communicate successfully in global virtual teams. Zemliansky writes, "Experiential learning is suitable for cross-cultural and virtual teams training since it creates conditions and environments that are most like those which will be encountered by learners in the workplace after they complete their training" [4]. Like previous scholars, Zemliansky acknowledges the strengths of experiential learning, and he applies them to current educational needs in international and online communication. In his review of literature, Zemliansky observes that there is a "remarkable degree of agreement among different fields" as to the value of experiential instruction [8].

Likewise, I have found an experiential approach to be particularly appropriate to learning skills in virtual and intercultural communication. If I were to teach these topics using non-experiential methods, I would likely have to isolate them and teach them in a fashion that relies on students' remembering lists of facts. Yu [18] points to the ineffectiveness of such a "factoid" approach to intercultural communication. Memorizing factoids doesn't foster any particular skill, and students are at risk of reaching an understanding of culture that is vague and based on contextless factoids that can result in an ethnocentric perception of culture [18].

When students actually work in global virtual teams, I don't have to isolate the teaching of intercultural and virtual communication, separating concepts and reducing context. Instead, students engage in a real experience where they interact with one another and their environment for a learning experience that fosters multiple skills. Similarly, Zemliansky and Herrington have used experiential learning projects to help students develop professional communication skills for cross-cultural virtual contexts [8, 9]. For example, Herrington and Tretyakov describe the Global Classroom Project as follows:

> The overall intent of the GCP is to study technical communication in practical application, and specifically, to develop effective cross-cultural, digital communication in subject-specific areas such as history, sociology, and political science. The assignments are centered on student collaboration. Students jointly develop print and digital products based on analyses of a content-specific topic. [9]

Thus, the goals of the GCP are to offer students the opportunity to develop both professional communication and subject-specific skills. Specifically, the GCP provided learning opportunities for graduate students in information design (United States) and graduate students in the humanities working on their English skills (Europe). Herrington and Tretyakov describe their GCP goal as "an experiment in cross-cultural, digital communication to provide a forum for experiential learning and a basis for communication research and analysis" [9]. Unlike the module I present in this chapter, Herrington and Tretyakov had a goal of eliminating instructor control. The module I present here emphasizes instructor involvement in order to attain a flexibility that allows any instructor in any field to construct global virtual teams in a given context of their choosing.

8.4 Teaching the Topic

Virtual team projects can provide powerful experiential learning opportunities in professional communication and engineering classrooms. Such projects provide an excellent method to better prepare students for a workplace where most of them will need to have both intercultural and virtual team communication skills (see Figure 8.2).

As Figure 8.2 illustrates, global virtual teams provide a real context for teaching and learning such skills as writing instructions or designing a website as well as improving professional competence in cross-cultural, team, and online communication skills. The

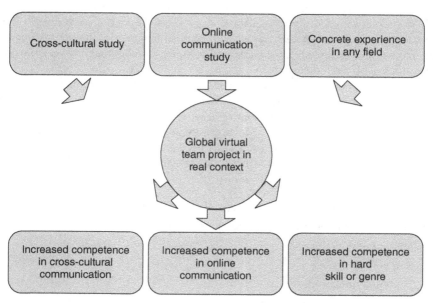

FIGURE 8.2. Experiential learning applied in global virtual team project. The rich learning context of global virtual teams supports improving competence in many skills simultaneously.

learning experience is rich and multi-layered, and no single skill could be learned as effectively outside a rich context such as this.

> Global virtual teams provide a real context for teaching and learning such skills as writing instructions or designing a website as well as improving professional competence in cross-cultural, team, and online communication skills.

8.4.1 The Module-Based Approach

Over the past two years, I have been using global virtual team projects in my classes and developing a virtual team teaching module that may be adapted across disciplines in many types of classrooms. Specifically, I have been constructing global virtual team projects between my professional communication students and students in universities around the world. For example, I have constructed virtual teams from among students in my classes and students in a tourism class in Armenia and students in business classes in Taiwan. With a collaborating international instructor (someone located in the other country), I give each team a common purpose consistent with a concept we are teaching such as researching and writing proposals.

Goals for the students in both my class and in the class of my international partner include increased virtual team, online, and international literacy as well as increased skills in professional communication, a specific genre, and writing for translation. Usually, the international partner class also wants to work on English language skills, so this item becomes an additional goal for such projects. Though goal priority is likely to be somewhat different in such cases (with English language skills predominant for some international partners), the experiential learning is rich enough to accommodate the priorities of all team members.

The flexible module I propose in this chapter is based on four teaching cases. I gathered data over a period of four academic semesters between spring semester 2010 and fall semester 2011. In each of these four semesters, I taught global virtual communication skills to junior and senior professional communication students as well as skills in writing instructions, proposals, and reports. I used an experiential model as described in the previous section and gathered the following artifacts during each of these four semesters:

- E-mail correspondence between myself and my partner instructors
- Student forum postings throughout the projects
- Student peer evaluations
- Student self-reflections
- Draft and final products
- Instructor reflections

In Sections 8.4.2 through 8.4.5, I provide background on each of the projects from these classes; then, in Section 8.5, I report on patterns I found in the experiences and the student/ instructor observations from all of the projects. Observations were shared both formally through interviews and reflection reports as well as informally.

8.4.2 Case 1. Class, Introduction to Professional Writing. Spring 2010. Face-to-Face Class Format. One Virtual Team Project—A Report. Partnered with Applied Foreign Languages Students at Ching Yun University (CYU), Taiwan

Background and Preparation This partnership was made possible by networking through Appalachian State University's Office of International Education and Development. As a component of this partnership, I travelled to Taiwan in 2009 and taught a class for Dr. Yi-chuan (Jane) Hsieh of the Department of Applied Foreign Languages at Ching Yun University (now Chien Hsin University of Science and Technology). Afterward, I contacted Dr. Hsieh and invited her to participate in a global virtual team project with me. We corresponded via e-mail to construct the project, and then we corresponded via e-mail each week throughout the project in order to work through problems. We acknowledged from the outset that this process was likely to be messy and have challenges such as the difference in language, time zone, and school calendar (to name but a few). However, with an experiential approach to instruction, there is much to be learned from contextual challenges including successes and failures. By the conclusion of the project, only two of seven teams had worked together effectively. However, due to the emphasis on process, all seven teams experienced learning, particularly increased proficiency in global virtual communication. To begin the project (a report), I shared with Dr. Hsieh, and then she with her students, a Western approach to report writing, documentation expectations, and formatting conventions.

Project Structure Dr. Hsieh was not teaching the same type of course I was, so we needed to be flexible. We decided that she would invite some of her best students who were interested in business English to participate in the project in exchange for extra credit. Seven CYU students volunteered. I split my class of 20 into seven teams, so each global team was composed of two to three Appalachian State University (ASU) students and one CYU student. Each team then designed one report.

In this first attempt, Dr. Hsieh and I made the credit low stakes as we didn't want the students to feel a great deal of pressure. (We used extra credit for Dr. Hsieh's students and no grade for how well the virtual team went with my students.) The final group product, the report, represented a significant percentage of their final class grade, but students could complete it successfully regardless of the "success" of the virtual team. For example, if a team experienced significant problems with language and understanding their teammates, they would have to adopt measures to mitigate the problem.

In one case, a team moved from a shared writing model to one where the CYU students provided an Asian perspective on the global report but did a limited amount of the writing and editing. This approach allowed the CYU students to accomplish their major goals of practicing English, working internationally, and becoming more familiar with Western business writing forms. My students could, in turn, accomplish their major goals of improving their competency in international and online communication as well as becoming proficient in the report genre.

Learning Goals While it is appropriate in an experiential learning environment to learn continuously within the context of a situation, Dr. Hsieh and I began with broad goals which we hoped to accomplish with this project. As for myself, I wanted my students to gain awareness and skill working in global virtual situations and a better understanding of writing for translation. In addition, I was specifically teaching the genre of report writing, so I wanted my students to gain skill in writing reports.

For Dr. Hsieh, the goals were somewhat different. She most wanted her students to have the opportunity to practice their English skills in realistic work situations. She also wanted to expose them to Western business writing. Notice, in both of our cases, we did not have a specific level of skill in mind; rather, we wanted to improve competency (students' ability to perform effectively).

Concrete Experience Working in teams made up of both ASU and CYU students, participants were to research and to write an informal report on guidelines employees should keep in mind as they work across cultures online. Dr. Hsieh's students were to research and contribute information on Taiwanese expectations in business communication, and my students were to share US expectations. All students were to contribute to the writing and editing associated with the project. Students were encouraged to select the technologies that they felt best supported their team and its goals. The most popular choices were e-mail, Facebook®, Skype®, and Google® Docs.

Modifying Behavior The fourth stage of the experiential learning cycle according to Lewin (refer back to Figure 8.1) is modifying behavior as a way to test the implications of the things students have learned. While students did modify their behaviors during the course of the project, and I will point to some of those, for the most part, behavior modification and testing would take place in future professional situations as well as in other classes. I also used some of the observations and concepts each team brought forward to inform my planning of future classes.

8.4.3 Case 2. Class, Business Writing. Fall 2010. Face-to-Face Class Format. One Virtual Team Project—A Report. Partnered with a Business Administration Class at CYU, Taiwan

Background and Preparation In my second attempt with conducting a global virtual team project, I found that Dr. Hsieh was not teaching a class that she felt was ready for this type of learning experience. We also agreed that the students on both sides of the project must receive similar credit. The non-equivalent credit our classes had received in the previous semester was a problem that affected optimal learning. After some networking, I found that Dr. Li-Hwa Hung in the Department of Business Administration and Graduate Institute of Business and Management at CYU was interested in this undertaking. Both my class in business writing and Dr. Hung's class in business administration were taught in a face-to-face format.

Project Structure My business writing class and Dr. Hung's business administration class worked together, both made up of junior- and senior-level undergraduate students. My class contained 17 students, and Dr. Hung's class contained 21. We planned to have students submit one report per team at the end with each of us grading our own students. We then placed the students into eight teams of four to five students each. Two ASU students were put into groups with two to three CYU students. Together they designed a report.

Learning Goals Learning goals remained the same as those from the previous semester.

Concrete Experience Working in teams made up of both ASU and CYU students, participants were to research and write an informal report on guidelines employees should keep in mind as they work across cultures online.

Modifying Behavior As with the previous semester, behavior modification was largely deferred, and testing of what was learned would take place in future professional situations as well as in other classes. However, students did modify behaviors in small ways throughout the course of the project.

8.4.4 Case 3. Class, Business Writing. Spring 2011. Hybrid Class Format. Two Virtual Team Projects—A Proposal and an Analytical Report. Partnered with a Tourism Class at Yerevan State Linguistic University (YSLU), Armenia, and a Management Class at CYU, Taiwan

Background/Preparation During this semester, I again worked with Dr. Hung at CYU to partner with one of her business classes. I also added a second virtual team project. This second project involved the drafting of a proposal and it required my students to collaborate with students in the tourism program at Yerevan State Linguistic University (YSLU) in Armenia. My Armenian teaching partner for this project was Ms. Annie Martirosyan, a YSLU instructor whom I met via one of her colleagues at YSLU.

Project Structure I had nine students in my course and taught it as a hybrid class. One third of our meetings were face to face, and two thirds took place online via ASU's Moodle site and Wimba web conferencing. Ms. Martirosyan had ten students in her class and taught it face to face.

For the YSLU project, students worked in teams comprising one ASU and one or two YSLU students. The students submitted one proposal per team, and we, the instructors, graded the proposals for our own students though we both gave comments on all drafts. Both instructors then suggested that the YSLU students edit and submit the proposals to an appropriate funding agency. During this same semester, my students collaborated with Dr. Hung's Human Resource Management class in Taiwan. Dr. Hung had 20 students in her class, and we constructed nine teams, each with two to three CYU students and one ASU student.

Two global virtual team projects in one semester provided a real challenge for my students, but they also had an excellent opportunity to learn by comparing their projects with students from two very different cultures. In creating this learning experience, my partners and I did not place the students from three nations into teams. Rather, my students completed one project with their Taiwanese partners (a report on Eastern influences on business), and they then completed a second project with their Armenian partners (a proposal for international travel support).

Learning Goals Learning goals remained the same as those from previous semesters.

Concrete Experience For the ASU/YSLU project, students researched funding opportunities and wrote proposals for international travel. For the ASU/CYU project, students researched and wrote a report on management strategies for organizations of the future given shifts in the relationship between East and West. To initiate these activities, my partners and I began by giving the students several articles on East/West managerial styles.

Modifying Behavior In this case, students thought they were going to be able to use what they learned in the first global virtual team project to increase the success of their approach to the second project. However, the students found that the learning was not generalizable given the two very different cultures with which they were working. I can best characterize student responses to the inability to generalize as surprised. That is, the students had approached the projects thinking of global virtual team situations as a somewhat homogenous experience, but soon discovered unexpected challenges and variables that affected interactions.

8.4.5 Case 4. Class, Business Writing. Fall 2011. Online Class Format. One Virtual Team Project—Analytical Report. Partnered with an Economics Class at CYU, Taiwan

Background and Preparation For this project, my online class at ASU partnered with a face-to-face class at CYU. Dr. Hung was not teaching a class whose English skills were ready for such a project in English. For this reason, I partnered with Dr. Jiann-fa Yan, Director of the International Cooperation Office at CYU, and his economics class.

Project Structure Because we knew from previous experience that the language barrier is great between CYU and ASU students, we organized this project differently. Dr. Yan had two visiting students at CYU, and both had native English-language skills. Dr. Yan split his class into two teams with each of the two native speakers of English leading a team. Both of these teams consisted of six to seven Taiwanese students.

I split my 14 US students into seven teams of two people each. I then assigned three to four of my teams to one of Dr. Yan's teams. In this model, CYU and ASU students did not write together on a single project. They instead served as primary resources for one another while each class worked on the same assignment, but did so separately (i.e., not as part of a common, international project team). In this case, the project was writing a report on US and Asian views on the global economy. My students served as resources for questions on the US views, and Dr. Yan's students served as resources for questions on the Asian views. The students, however, wrote in teams that comprised exclusively of individuals from their own classes.

Learning Goals Learning goals remained the same as those from previous semesters.

Concrete Experience Students researched and wrote a report comparing the US and Asian views on the global economy. As with previous projects, the students participating in this project could choose the technologies that best supported their team and goals. In undertaking this project, one group of students immediately set up a Facebook group to coordinate the exchange of ideas and information.

Modifying Behavior As with other semesters where I hosted a single, global virtual team project, behavior modification was largely deferred, and testing of what was learned would take place in future professional situations as well as in other classes.

8.5 Observations/Reflections/Theory Development for All Classes

In this section, I report on the patterns that emerged from among the four individual projects described in sections 8.4.1 through 8.4.5. As students and instructors immersed themselves in the projects, we began to observe and reflect on the experience as is consistent with the second stage of Lewin's experiential cycle (see Figure 8.1). Based on what we observed, we (students and instructors) began to reflect and develop theories that might explain these observations.

In some cases, students were able to reflect, generate a theory, and test that theory within the context of their projects. In other cases, theories would have to wait to be tested in future projects. Table 8.1 presents an overview of observations, reflections, and related theories that students and instructors articulated via interviews and reflection reports.

As Table 8.1 indicates, students and instructors had much to say about what they observed and what theories might explain those observations. All participants found the experience difficult and unpredictable but rewarding. They also found that a strong articulation of purpose and consistent communication (with much feedback and few delays) were important to team success. Participants recognized that technology preferences, the need for social interaction, and perceptions of time varied with culture. Overall, students gained an increased appreciation for cultural differences and the impact of language.

TABLE 8.1. Student and Instructor Observations, Reflections, and Proposed Theories

Category	Observation/Reflection	Assimilation
Overall success	Instructors commented that the projects benefitted students in ways consistent with our original goals. Overall, approximately 60% of teams reported that the projects went well while 40% reported that they did not go well. However, approximately 95% of students said that the projects were a valuable experience and would participate again. YSLU students most valued the opportunity to communicate in English (writing and speaking), work in teams and assert their opinions, work with deadlines, write "official" documents, broaden cultural perspectives, and establish international relationships. "The students were so enthusiastic about it all that they did not even bother about the time gap (a 9-hour difference between Armenia and North Carolina) to chat!" (Ms. Martirosyan, YSLU).	The goals of one group of students are not (nor need they be) the same as the other groups'. Each group will have a different learning experience, but all groups will experience learning that increases their global and virtual communication competence. "Almost all the points we came up with in our report we based off the struggles we faced in our virtual team." "This project really made me step out of my comfort zone. Generally, I tend to be very individualistic in group projects. Forcing myself to communicate with the Taiwanese as well as [my ASU partner] helped me realize the value in teamwork." "This experience is good practice for business communication—writing to multiple audiences, writing clearly and concisely."
Purpose	Students were willing to work hard but often felt confused. They often reported that their partners were unsure of what to do at points in the project.	Everyone needs a strong, common understanding of the project including detailed guidelines and schedules. "Crossing the communication barrier is the most essential part to having an effective virtual team. In order for the assignment to get accomplished most proficiently, all sides must have a firm grasp on what it is that is assigned and their role in producing the finished product."

TABLE 8.1. Student and Instructor Observations, Reflections, and Proposed Theories

Category	Observation/Reflection	Assimilation
Initial time	Students need more time.	Allow more time for initial contact before establishing a business relationship. Include planning for punctuality. "It is very hard for them [Taiwanese students] to work with people they do not know personally. I will know that next time; communication between parties must start much earlier than when the project starts."
		"My own research, as well as our group's communication suggests that this [establishing relationship] foundation is critical, as its conception shows that each party involved has the utmost respect for each other's culture; this affirms the mutual determination to synergize and complete tasks on time."
		"A long, relaxed period of time may help reduce challenges with language, culture, and time zones."
Technology	Most students preferred e-mail for communication as the most reliable form of communication even though they tried several technologies.	Instructors should require the use of particular tools, in this case Skype and Google docs, in order to "enable better development of both oral and written language."
	There was active debate about which technologies were most useful. Besides e-mail, Skype and Facebook were favorites although CYU students usually did not care for Skype.	"Use of Facebook only promotes too much social conversation."
		"Though we would have liked to communicate directly with our team members via live chat, the time difference (approximately 12 hours) did not enable us to participate in this form of communication. This forced us to rely on e-mail and Facebook postings alone, which did not seem to be as effective due to the turnover rate of communication."

TABLE 8.1. Student and Instructor Observations, Reflections, and Proposed Theories

Category	Observation/Reflection	Assimilation
Feedback	Students experienced many problems receiving responses—the correspondence was often unclear.	"The low-achievers [in English as a second language] were intimidated to read e-mail messages from their American peers since the messages were mostly in the textual form with a lot of English unknown words and sentences were hard to comprehend" (Dr. Hsieh, CYU).
Types of communication	In many teams, CYU students preferred more social communication than did ASU students.	Social communication may have been a strong preference for students who were the least comfortable with English as a second language. "The English high-achievers benefited from the project because they were inspired by their American partners. They tried to think aloud, actively search for information, evaluate what they found, and systematically organize their report. As for the English low-achievers, they were actually fond of sharing Taiwanese cultural heritage with their American partners. They were excited to talk to their American partners through MSN using simple English" (Dr. Hsieh, CYU).
Delays	It was difficult for ASU students to contact CYU partners, or responses were unpredictable/many delays.	Delays in hearing back were sometimes an overwhelming issue. ASU students hypothesized these were due to difficulty with language.
Perceptions of time	Perceptions of punctuality were very different. ASU students observed that CYU partners were late on deadlines more often than ASU students. CYU students commented that responding in a timely fashion seemed very important to ASU students.	One YSLU student suggested that setting regular hours for communication would help them to schedule their days. ASU students hypothesized that sense of time was very different between themselves and students at CYU.

TABLE 8.1. Student and Instructor Observations, Reflections, and Proposed Theories

Category	Observation/Reflection	Assimilation
Culture	Many students commented on gaining a better appreciation of the other culture. "CYU students seem shy but not so shy as you get to know them."	"Confusion offers some of the greatest opportunities for learning. As you work through misunderstanding, don't get frustrated as this is unproductive." "Languages and cultures really do allow for different views on issues." Some missed deadlines could be due to cultural differences. "I will take a more collective approach in the future when working with Asian colleagues." ASU students say they will be more persistent in communicating rather than waiting. "It may be necessary to be more demanding than you are comfortable with." "We learned that trust and relationship are more important to Asian partners than U.S. partners."
Language	The language barrier varied widely by team and could be a great challenge. CYU students appreciated it when ASU students adapted their English for easier understanding and were concerned about their English skills. Dr. Hung's students were slow with their English and, therefore, felt they were less proactive on the project. They assumed more of a resource role.	"Language is a challenge but this helps us learn about writing for translation." "Learning to write in simplified English is valuable." "A language liaison may help. Language/understanding was challenging." ASU students felt that lack of confidence in English skills (by CYU students) may have caused problems with frequency of communication. "... must be more careful with word choice when the language gap is bigger." Relying more on asynchronous communication may help. Language ability links to leadership opportunities.
Credit	Extra credit for CYU students didn't work.	All students must have similar levels of investiture—receiving similar levels of credit as it impacts them. The levels of commitment must be the same on both sides, or roles must be adjusted. For example, students in one class may act as resources rather than partners.

(For quotations, instructor names are noted with their permission; however, student names are not noted for privacy reasons. Observations and assimilations without quotation marks are provided by the author based on multiple sources of input from the artifacts.)

8.6 Global Virtual Team Teaching Module

Students and instructors who have participated in these global virtual team projects have found the experience worthwhile though the process is often messy and uncomfortable. Students have described their experience as "ground-breaking" and "enlightening." One student described the experience as "exciting in theory, yet disappointing in practice"; this same student, however, went on to describe significant learning. This student's conflict might have been due to student perceptions that projects have to go smoothly for success to be achieved. In experiential projects, and in the world of work, we know that the knowledge gained through mistakes, failure, and *things not going quite right* is an important form of success. It makes us more skilled in the future. I have also received e-mails from students once they have graduated saying that the global virtual team experience helped them get a foot in the door with employers.

> Students and instructors who have participated in these global virtual team projects have found the experience worthwhile though the process is often messy and uncomfortable.

8.6.1 Instructor Development

Even as students learn experientially in global virtual teams, I have been participating in an experiential learning process myself regarding the use of these teams. During the four full semesters of observing and reflecting on the use of virtual teams to teach concepts of professional, virtual team, and international communication, I have begun to assimilate these observations into theory—a basic module that may aid other educators in launching such projects. I provide an overview of this module in Figure 8.3.

Using the afore-mentioned module, educators and trainers can construct teams from among their students and students at universities or facilities in other countries. The module is flexible and can be used in any type of course (in industry or education) as well as between different types of classes, different levels of classes, and different branches of an organization. Flexibility is essential because educational needs change, as do the number of students and English competency. In addition, flexibility allows this module to be adapted successfully in corporate training or university courses.

8.6.2 Foundations for Implementation

In the following subsections, I provide a more detailed description emphasizing *strategies for success* in each of the components of the module.

Establish Instructor Partnerships Make contact with other interested faculty using international offices at your organization and your own network. While this is often the first question that others ask me about the "how to" of the project, it is, perhaps, the easiest part of the process to perform. Opportunities abound as universities and other types of organizations are increasingly global in nature.

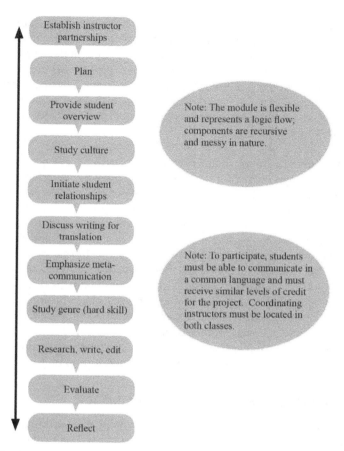

FIGURE 8.3. Module of critical task categories supporting global virtual team projects. Using this module, educators and trainers can construct projects among their students and students at other international sites.

Plan The instructors of each class should now plan an appropriate project. This planning needs to be thorough employing much of the same process that will be taught to students: agree on the project goals, structure, pedagogy, and credit value. Common purpose(s), understood by all, is critical [19, 20]. In addition, instructors should allow a generous amount of time for the project, increasing the amount of time allocated as shared language skills decrease. During this time, instructors should also decide how the individual teams will be structured. Just as I will advise my students to communicate explicitly in order to reduce potential miscommunication [20,21], I must be very explicit in planning with my international colleagues—agreeing to communicate often.

Provide an Overview for Students Providing an overview of a project is always a sound approach because people perform tasks better when they understand the scope and goals before they begin to work. This factor becomes especially important in launching a global virtual team project where there are many challenges corresponding with the many levels of learning. Give students a clear picture of the scope of the project and articulate the value of both successes and failures to the learning process. (Note: I do not evaluate students on the perceived success of their teams. I evaluate them based on their effort and the quality of the final product. I make this very clear at the beginning of the project so that they can feel free to experience and experiment with communication, technology, and so on.)

Study Culture Learning about other cultures and increasing students' cultural competence is a core opportunity for such virtual team projects. Nevertheless, instructors should not be discouraged from attempting such projects because they do not have time available for a deep study of culture. Students will increase their cultural competence during the project regardless of the time set aside for the explicit study of culture.

The study of culture can be addressed in relatively short time or over a long period of time as is appropriate to the context of the participating classes. In addition, students can increase global competency by recognizing that cultural distinctions are not limited to national culture. As Hunsinger notes, "…traditional definitions of culture are often unstable and even unreliable in online environments… technical communicators might find it more effective to localize online materials based on a broader set of cultural and noncultural contexts (e.g., legal, economic, and technological) that influence an online audience's behavior [22]. While Hunsinger is addressing website communication, this approach to culture seems also applicable to the relationships in online teams.

8.6.3 Making Contacts and Sharing Experiences

In my classes, I might spend a day of class time discussing national cultural characteristics with my students, or I might invite in a guest lecturer who is very familiar with the culture with which we are working. In some cases, I might assign a few readings and have students post to a discussion forum, or I might use a combination of all of these approaches. In essence, the methods I use depend on the topics of the classes involved, the assignment students are given, and the availability of resources related to a given culture.

Put Students in Touch with One Another After my international partner and I have provided an overview to students, we put the students in touch with their teams by providing them with the e-mails of their teammates. For these projects, students are not required to use e-mail as part of their communication technologies; but all students have e-mail, so it is a good place to start. My partners and I also give students ideas for technologies they might wish to use for their projects, but the final choice is theirs to make.

Some favored technologies for these projects have been e-mail, Skype, Facebook, and Google Docs. Depending on the context (such as language ability, school calendar, class topic, and so on), we try to allow a generous amount of time for students to get to know one another. As is illustrated in the student observations in Table 8.1, this is one of the bigger challenges to these activities. Students need time to establish relationships, and, of course, this is more important in some cultural contexts than in others. As noted by Walter [23], communication based on trust is slower to start in online communication than in face-to-face communication

An increase in cultural distance and language correlates with an increased need for this relationship-building time at the beginning. In the case of instructors who teach on semesters or quarters, differences in school calendars can make finding longer amounts of time very difficult. I have begun to stretch this time as far as possible by having my students begin this process while working on other class projects.

Discuss Writing for Translation At some point in the project process, it is important to discuss writing for translation. I often do this soon after students have opened informal discussions with their international partners, for it is at this point that they begin to experience some of the challenges in language directly. One short exercise I use during this time is to have students use a machine translator to translate some phrases they need when interacting in this partnership. Table 8.2 provides example results of some student efforts to use these technologies in their international project teams.

Through such activities, students quickly realize that machine translation is less than perfect for conveying their meaning across languages. Between this exercise, what students are experiencing in their partnership correspondence, and class discussion, students begin to see the large role that language plays in the success of international projects, and they begin to learn how they might adapt their own writing so that it is more easily translated.

Emphasize Meta-Communication During the time that my students are beginning to establish a relationship with their international partners and are studying culture, I begin to encourage them to include *meta-communication* in their international conversations. I use meta-communication to refer to *communicating about communication.*

TABLE 8.2. Example of Communication Using a Mechanical Translator

Initial Communication	Translated to Chinese	Translated from Chinese Translation to English
Did you receive the e-mail?	你收到的电子邮件吗?	You receive the e-mail?
Have you edited and revised the final draft?	你编辑和修订草案定稿?	You to edit and revise the final draft of?
We will be sending you this document upon completion.	我们会向您发送完成后?本文档。	We will send you after the completion of this document.
Did you read the last article we sent?	你读过上一篇文章中?我们派出了吗?	Have you read the previous article, we have sent you?

The information in this table demonstrates the limitations of machine translation.

Through my own research in global virtual teams [5] and research preparing online student cohorts to work in 3D Worlds, I have both identified and tested the significant role that meta-communication can play in effective online communication. Thus, during the earlier stages of virtual projects, I encourage my students to discuss communication preferences with their international partners. They should, for example, discuss the technologies they prefer, the times of day that are best for working together, the concept of punctuality, what project components they most want to work on, and so on. This conversation cannot begin until the relationships are underway, but it does need to take place before the work on deliverables begins.

Study Genre Once the informal communication is well underway, I and my international partner can begin to introduce the project; for me, this is always some genre of professional communication as these are the classes I teach. Two of my favorite genres for this project are *reports* and *proposals* as they are of value in all disciplines. Thus, whether or not my students are working with tourism students in Armenia or business students in Taiwan, all students can benefit from learning to write effective reports and proposals. Some of the topics I have used are described earlier in the chapter in the "concrete experience" sections.

A primary challenge in teaching genre and launching the project is that my colleague who is instructing the international class is usually not an authority on genres in professional writing. Thus, addressing this component of the module often falls to me. I prepare PowerPoint® presentations and scan book sections that I send to my colleagues so that they might prepare their students for discussions of this topic. I willingly answer questions, and I acknowledge that the level of skill I expect from my students (in a professional writing class) will be higher than for students in, for example, a tourism class. Nevertheless, all students become more competent in the assigned genre.

Research, Write, and Edit At this point, students engage in the traditional writing cycle with input from me and my partner instructor. Depending on the language gap, some students might do more writing than others; however, both my students and their international partners serve as primary resources for one another representing their own cultures. For example, if my students are working with an undergraduate class in Taiwan whose English is not yet advanced, my students might do most of the writing while the Taiwanese students provide resource material.

Evaluate Students turn in their final projects for grading, and I grade my students' work while my international partners grade their students' work. This is critical because I will always be grading my students most heavily on effort in the team and quality of the professional communication product. Quite often, my international partners will be grading their students on products that demonstrate English usage in addition to other competencies that are most important to them within the context of a given course.

Though we may be evaluating different competencies and weighting them differently, the overall project must have equivalent value for both my students and their international partners. It is in this way that all students are similarly vested and, thus, are willing to expend similar levels of effort for these projects. As was discussed previously, early in my working with virtual teams on one project, Dr. Hsieh assigned the project as extra credit while the project carried a value of 15% of the class credit for my students. Because her students were not similarly vested in the project, they did not feel any pressure toward goals and deadlines; in turn, my students felt increasing pressure.

Reflect Reflection is an essential part of the experiential learning process. It is during reflection that students move from observing the doing to thinking about what their observations may mean. I encourage students to reflect during class discussions throughout the project, in asynchronous online forums, in their peer evaluations at the end of the project, and in a letter of reflection at the end of the project. Students often acknowledge the challenges and discomfort they felt but then move into perceptions of their own learning. It is at this stage of the project that I often feel the greatest sense of having facilitated a meaningful learning experience.

8.6.4 Experiences Applying Ideas and Approaches

During one of the semesters I described earlier in this chapter, I assigned two of these projects to a single class. There was great value in doing two virtual team projects in one semester although it was quite a challenge for students. Students immediately noticed that issues with time such as delays and the clarity of language were quite different when comparing their two projects. One student wrote in the discussion forum, "Guys, I think this is the difference between the West and the East. We Westerners put great weight on deadlines and time while the Easterners are more relaxed." Their first project had been done with students in Armenia who seemed to have more Western views than the Taiwanese students with whom they collaborated on the second project. Further reflection from one student designed this quote, "At the beginning of each project I undertake with an international partner, it is important that I start at square 1, and make sure I handle all elements of meta-communication appropriately. The communication with CYU was nothing like the communication with Yerevan."

8.7 Conclusion

Global and virtual team skills are prized in the field of engineering as organizations increase their global reach. Skills that ensure the success of such teams are not intuitive and should be taught and practiced in settings that do not affect an organization's product. The best model for teaching these skills may well be an experiential one where students (whether or not they are practicing professionals) practice the skills in real, low-stakes situations. As described in this chapter, I have been using an experiential learning model for several years to teach global virtual team skills. Students perform in real global virtual teams, observe and reflect on the experience, and assimilate theories to help explain successes and failures. They can then carry this experience forward into their professional lives.

Global virtual team projects, as I have described them here, connect students with international opportunities. Participating instructors can globalize their classrooms in real ways, inexpensively and realistically. Students gain real international and virtual team experience within a classroom setting without additional expense to a program. They practice speaking and writing English, writing for translation, working in a virtual team, and collaborating internationally. In a professional writing collaboration, they also gain skills in a genre. Most students are engaged by this concrete experience and find that they are more prepared to work internationally and online in the future.

References

1. J. Eison, "Teaching strategies for the twenty-first century," in *Field Guide to Academic Leadership*, R. M. Diamond and B. Adam, Eds. San Francisco, CA: John Wiley & Sons, 2002, pp. 157–173.
2. S. Sarker, S. Sarker, and C. Schneider, "Seeing remote team members as leaders: A study of US-Scandinaivan teams," *IEEE Transactions on Professional Communication*, vol. 52, no. 1, pp. 75–94, 2009.
3. A.-F. Rutkowski, D. Vogel, M. Van Genuchten, and C. Saunders, "Communication in virtual teams: Ten years of experience in education," *IEEE Transactions on Professional Communication*, vol. 51, no. 3, pp. 302–312, 2008.
4. R. Raju, "Intercultural communication training in IT outsourcing companies in India: A case study," *IEEE Transactions on Professional Communication*, vol. 55, no. 3, pp. 262–274, 2012.
5. P. E. Brewer, "Miscommunication in international virtual workspaces: A report on a multicase study," *IEEE Transactions on Professional Communication*, vol. 53, no. 4, pp. 1–17, 2010.
6. G. R. Berry, "Enhancing effectiveness on virtual teams: Understanding why traditional team skills are insufficient," *Journal of Business Communication*, vol. 48, no. 2, pp. 186-206, 2011.
7. B. A. Olaniran and D. A. Edgell, "Cultural implications of collaborative information technologies (CITs) in international online collaborations and global virtual teams," in *Handbook of Research on Virtual Workplaces and the New Nature of Business Practices*, P. Zemliansky and K. St.Amant, Eds. Hershey, PA: Information Science Reference-IGI Global, 2008, pp. 120–136.
8. P. Zemliansky, "Achieving experiential cross-cultural training through a virtual teams project," *IEEE Transactions on Professional Communication*, vol. 55, no. 3, pp. 275–286, 2012.
9. T. Herrington and Y. Tretyakov, "The global classroom project: Troublemaking and troubleshooting," in *Online Education: Global Questions, Local Answers*, K. C. Cook and K. Grant-Davie, Eds. Amityville, NY: Baywood, 2005, pp. 267–283.
10. K. St.Amant, "Online education in an age of globalization: Foundational perspectives and practices for technical communication instructors and trainers," *Technical Communication Quarterly*, vol. 16, no. 1, pp. 13–30, 2007.
11. M. C. Paretti and L. D. McNair, "Communication in global virtual activity systems," in *Handbook of Research on Virtual Workplaces and the New Nature of Business*, P. Zemliansky and K. St.Amant, Eds. Hershey, PA: Information Science Reference-IGI Global, 2008, pp. 24–38.
12. L. Meng and R. Schafer, "An international virtual office communication plan," in *Virtual Workplaces and the New Nature of Business Practices*, P. Zemliansky and K. St.Amant, Eds. Hershey, PA: Information Science Reference-IGI Global, 2008, pp. 555–568.

13. E. Saatci, "Problem-based learning in an intercultural business communication course," *Journal of Business and Technical Communication*, vol. 22, no. 2, pp. 237–260, 2012.
14. K. Lewin, *Field Theory in Social Science*. New York: Harper & Brothers, 1951.
15. J. Dewey, *Experience and Nature*. New York: Dover Publications, 1958.
16. J. Piaget, *Psychology and Epistemology*. Middlesex, England: Penguin Books, 1971.
17. D. A. Kolb, *Experiential Learning: Experience as the Source of Learning and Development*. Englewood Cliffs, NJ: Prentice Hall, 1984.
18. H. Yu, "A study of engineering students' intercultural competence and its implications for teaching," *IEEE Transactions on Professional Communication*, vol. 55, no. 2, pp. 185–201, 2012.
19. L. F. Thompson and M. D. Coovert, "Understanding and developing virtual computer-supported teams," in *Creating High-Tech Teams*, C. Bowers, E. Salas, and F. Jentsch, Eds. Washington D.C.: American Psychological Association, 2006, pp. 213–241.
20. J. Lipnack, *Leading Virtual Teams: Managing People at a Distance*, ed. New York: American Management Association, January 25, 2013.
21. A. H. Reed and L. V. Knight, "Effect of virtual project team environment on communication-related project risk," *International Journal of Project Management*, pp. 1–6, 2009.
22. R. P. Hunsinger, "Using global contexts to localize online content for international audiences," in *Culture, Communication, and Cyberspace*, K. St.Amant and F. Sapienza, Eds. Amityville, NY: Baywood, 2011, pp. 13–37.
23. J. B. Walther, "Computer-mediated communication: Impersonal, interpersonal and hyperpersonal interaction," *Communication Research*, vol. 23, no. 3, pp. 3–43, 1996.

Educational Contexts

9

Strategies for Developing International Professional Communication Products

Helen M. Grady

Mercer University

In a global business environment, professional communication products must meet the significant demands of different cultures and languages. To create a viable product in the international marketplace, we need to understand how to balance business needs with culturally diverse users' needs. This chapter presents strategies for teaching students how cultural variables (beliefs, attitudes, and values) shape the communication process involved in such design practices, and it includes examples of student assignments and student responses to those assignments. The chapter also contains specific examples of the practices that professional communicators need to consider when internationalizing and localizing communication products.

9.1 Introduction to International Technical Communication

Mercer University's Master of Science in Technical Communication Management (MSTCO) is geared toward practicing technical communicators, who have an average of 10 years' experience when they begin the program. Increasingly, these graduate students report that they are working in a multicultural environment in which both coworkers and customers come from a variety of cultural backgrounds. Students also report that more of their employers are participating in the global marketplace, and this factor is

---

Teaching and Training for Global Engineering: Perspectives on Culture and Professional Communication Practices,
First Edition. Edited by Kirk St.Amant and Madelyn Flammia.
© 2016 The Institute of Electrical and Electronics Engineers, Inc. Published 2016 by John Wiley & Sons, Inc.

driving the requirement to create information products in different languages. Thus, current marketplace practices have created a distinct professional need to understand and work within international and intercultural environments.

To help students meet this demand and be prepared to manage communication products and processes involving different cultures and languages, the faculty at Mercer created a course in International Technical Communication. The purpose of this course is to examine how cultural beliefs, attitudes, and practices shape communication processes. To achieve this objective the course examines the theory, research, and practices that technical and professional communicators and engineers need to consider when internationalizing and localizing communication products. The course focuses on strategies for communicating in a culturally diverse workplace and for preparing communication products for international markets.

This chapter presents an overview of the topics covered in the International Technical Communication course. These course topics include cultural models, rhetorical patterns of different cultures, internationalization and localization issues, factors to consider in translating communication products, and graphic design challenges for international audiences. The chapter also contains detailed information on assignments that require the students to convert theory into practice by creating two products (an oral presentation and a feasibility report) that are specifically related to their own workplace. Included in this overview of the class are grading rubrics for class assignments, student feedback on the relevance of the assignments and a list of assigned readings used in the class.

9.2 Review of the Literature

This review of the literature includes foundational theory as well as practical applications of the theory in the areas of

- Intercultural communication
- Cultural theories and models
- Internationalization, globalization, and translation

The faculty selected these areas because they are directly related to the course assignments.

9.2.1 Course Texts

Graduate students in the MSTCO program are primarily practitioners, many of whom work for international companies and therefore have a need to communicate with customers and coworkers from a variety of cultures. There are, in turn, a number of excellent texts on intercultural communication practices and strategies instructors can draw on when developing a course that meets the needs of these students. These texts include

- *Basic Concepts of Intercultural Communication* (1998) by M.J. Bennet
- *The Sage Handbook of Intercultural Competence* (2009) by D. Dierdorff

- *Intercultural Competence* (1998) by M.W. Lustig and J. Koester
- *Intercultural Communication in Contexts* (2004) by J.N. Martin and T.K. Yakanama
- *Managing Cultural Differences* (2011) by R.T Moran et al.

Each of these texts deals with some aspect of working with members of different cultures.

9.2.2 Growing Interest in the Field

Moreover, the increasing importance of intercultural communication to our field in general is evidenced by the number of special journal issues that focus on this topic. For example, *Technical Communication* has devoted four issues to various aspects of communicating in global environments (May 1999, November 2001, May 2006, and August 2011), while the *IEEE Transactions on Professional Communication* has published special issues related to global professional communication practices (March 2004, June 2007, September 2011, June 2012), as has the *Journal of Business and Technical Communication* (July 1997 and July 2010).

> An important element to understanding how to communicate with members of other cultures is an understanding of the various models of culture.

An important element to understanding how to communicate with members of other cultures is an understanding of the various models of culture. Hofstede's [1] seminal work in this area is a good starting point for examining such models, although his theories on the dimensions of culture have been criticized by McSweeney [2] and Signorini et al. [3] as lacking in empirical evidence, oversimplifying cultural differences, and treating culture as static, as have Hall's [4] theories on high- and low-context cultures by Cardon [5]. An excellent synopsis of the various theories on culture can be found in Gould [6]. Additionally, many authors have examined how culture impacts various aspects of communication such as interface design [7–9], discourse pragmatics in Latino cultures [10], patterns of communication in South American work places [11], and cultural considerations for communicating in China [12].

9.2.3 Fundamental Processes and Practices

Because many companies create products that are being distributed internationally, or will be in the future, an understanding of what is involved in localizing or globalizing an information product is becoming mandatory. Schaler, for example, estimates that 95% of all software that is localized originates in the United States. He believes the growth in localization is driven by the fact that "While advances in technology, trade and politics have made the world a smaller place, politics and business remain essentially local. People prefer to communicate in their own language and to conduct business in their own language" [13, p. 2]. These factors have given rise to a range of industries and practices designed to create products for culturally diverse audiences.

The Localization Industry Standards Association (LISA) defines localization as the actual adaptation of the product for a specific market. It includes translation, adaptation of graphics, adoption of local currencies, use of proper forms for dates, addresses, and phone numbers, and many other details. Internationalization, in contrast, is the process of designing a culturally-neutral product that can be localized, while globalization is a business process that includes the technical, managerial, personnel, marketing, and other business decisions needed to facilitate localization. (Unfortunately, LISA was dissolved as an organization in 2011, but their materials are still available at http://web.archive.org/web/20110101184308/http://www.lisa.org/). These practices, moreover, have gained increased attention and importance within technical and professional communication.

9.2.4 Connections to Professional Communication Practices

Several authors have examined the connections between localization and technical and professional communication practices in various organizations and industries. Related research into this area includes Walmer's [14] description of translation and localization processes for medical equipment documentation, Yli-Jokipii's [15] study of Finnish and English websites, and Sapienza's [16] examination of websites for Russian immigrants. More recently, Sun [17] has analyzed the use of cell phone text messaging in China, and St. Germaine-Madison [18] has examined the localizing medical information for Spanish speakers in the United States.

For technical communicators, a particularly important aspect of localization is translation. Melton provides an excellent review of translation theories as they relate to technical communication practices. He argues that translation is not "just the act of transferring or even recreating information, but also a holistic process of building relationships in which linguistic and cultural factors are inseparable" [19, p. 211] Other authors have provided guidance in preparing documents for translation [20–22]. Increasing recognition on the importance of translation skills is evidenced by the number of authors advocating for including training in translation in US schools [23, 24].

Research in technical and professional communication thus indicates the field is changing to become more global in nature. As a result, today's technical and professional communication students will need to acquire a range of skills and knowledge needed to work successfully in this new context. Technical communication instructors, in turn, need to develop the kinds of activities, assignments, and classes that allow students to develop such skills and abilities. Yet creating such items is no simple feat. For this reason, instructors in technical and professional communication can benefit from examinations of how others have created educational materials and focused courses that examine the connections between culture and technical and professional communication.

9.3 The International Technical Communication Course

The International Technical Communication course is part of Mercer University's MSTCO program, which is housed in the University's School of Engineering. Graduate students enrolled in the program are primarily from the United States and are all practicing technical communicators who have a minimum of 3 years' experience in the field (the average work experience for students is usually 10 years or more). The MSTCO degree is designed to prepare these individuals for leadership roles within their technical communication organizations.

9.3.1 Course Focus, Goals, and Objectives

To achieve this goal, the degree combines academic theory and research with best practices from industry. The program thus parallels the educational philosophy in the School of Engineering's other graduate programs: to provide quality education to working professionals. In fact, many engineering students minor in technical communication management in order to complement their graduate engineering degrees. The program is entirely online and is delivered by a combination of asynchronous (via the Web) and synchronous (via weekly 90-minute, face-to-face WebEx sessions) methods.

Within the context of the overall program, the educational objectives for the International Technical Communication course ensure that by the end of the course, students will be able to

- Explain how cultural factors affect the communication process with international audiences
- Describe the strategies of internationalization, localization, and globalization
- Make effective decisions about visual design for international audiences
- Identify management issues involved in developing effective international technical communication products

To achieve these learning objectives, the students complete an extensive set of readings, including two required books:

- *Usability and Internationalization of Information Technology*, by Nuray Aykin (ed.), 2005
- *Cultures and Organizations: Software of the Mind*, by Hofstede and Hofstede, 2005

The students also complete several assignments that focus on culture and communication in technical and professional communication contexts.

To examine these ideas on culture and communication, the class meets synchronously each week for 90 minutes via WebEx®. Prior to these synchronous discussions, students are required to post discussion questions or ideas to a specific area on a class Blackboard® site. Topics for discussion during the synchronous meeting are chosen from these questions. Each student is also responsible for leading the discussion during the synchronous meetings, and the role of discussion leader rotates among the students. In addition, the instructor posts questions related to the weekly topic to an asynchronous discussion board on the class Blackboard site, and students are required to contribute to a threaded discussion on a particular topic. Through this mixed approach, students are able to use different forums to examine an issue in both the short term (i.e., synchronous discussions) and over time (e.g., via the asynchronous discussion board posts).

9.3.2 Course Content

At the beginning of the course, the students assess their cross-cultural communication skills using Goodman's [25] self-assessment exercise. The exercise consists of 20 questions, and students rate their responses on a scale of 1–5. These results are always an eye-opener for US students, as most of them are surprised at how poor their cross-cultural communication skills are. The students and the instructor then discuss why it is important for technical communicators and engineers to have such cross-cultural communication skills. During this process, the importance of such skills are reiterated both in Hoft's [26] introduction to a *Technical Communication* special issue on international technical communication and Hayhoe's [27] *Technical Communication* editorial on the globalization of the profession.

The instructor then introduces students to various models of culture. In so doing, the instructor draws from Gould's [6] chapter on synthesizing the literature on cultural values as it is a particularly useful review of the cultural theories of Hofstede, Hall, Victor, Trompenaars, Triandis, Gudykunst, and others. The students also spend time comparing Hofstede's "onion" theory of how cultural differences are manifested with the iceberg metaphor for culture presented in Hoft [28]. Through these discussions, students continue to deepen their understanding of how cultural values shape various forms of communication by reading a variety of articles [2, 7, 9, 10, 12, 29, 30].

Next, the class examines the impact of culture on rhetorical patterns. To do so, students read Connor's [31] article, which provides a good overview of the field of contrastive rhetoric and notes its value in helping individuals understand different rhetorical patterns. Students also read several other articles [32–35] that examine different cultural preferences, including Campbell's [36] article, which provides students with a good, brief overview of Kaplan's styles of paragraph development as well as presents some interesting examples from international business communication. Students also read Wang's [37] article that explores the effects of the flattening of the world on culture and specifically on contrastive rhetoric.

> No study of internationalization and localization would be complete without an examination of the translation processes and translation issues.

In order to study of globalization, internationalization, and localization, the students compare the approaches of Hoft [28] and Aykin [38]. Hoft views globalization as the process of creating a product that is universally intuitive and can be used by many cultures. Aykin, in contrast, uses the Localization Industry Standards Association's (LISA) definition of globalization, which views localization as a business process that includes the technical, managerial, personnel, marketing, and other business decisions needed to facilitate localization. In examining these ideas and perspectives, the role of the instructor often becomes that of a guide who helps students come to an agreement on a common definition for these terms. (Doing so makes further discussion of these topics easier.)

From this point, students read and discuss McCool's [39] case study on how localization efforts can fail when cultural values are not considered. While there are many articles in the literature on superficial versus radical localization, this particular reading by McCool stands out in terms of teaching students, for it makes a good connection between cultural dimensions and radical localization. Thus, it is a good example of relating theory to practice. So is Sun's [17] article in which she argues for more radical localization, although she does not use that specific term in making her argument.

No study of internationalization and localization would be complete without an examination of the translation processes and translation issues. To examine these items, students read Hoft's [28] very practical discussion of translation strategies and issues. They must also read DePedro's [40] article on translation theories and are asked to pay particular attention to the concept of linguistic and cultural untranslatability. As translation quality is also an important topic, students also read House [41].

Because the graduate students are employed full time, it is also important to introduce them to the practical issues involved in translation. To do so, the students read articles by Alternero [42], Walmer [14], and Akis and Sisson [43]. In both the synchronous and asynchronous discussions of these readings, the students examine how various industries have tried to make their documentation more user and translation friendly by applying various approaches to clear writing.

The semester then concludes with a review of graphic design issues as they relate to international audiences. To examine such topics, students read Horton's [44] very pragmatic approach to internationalizing and localization graphics as well as Moore's [45] discussion of the cultural implications of graphic design, and Wang's [46] analysis of cultural preferences for visuals in scientific documents. The instructor also asks students to examine several of the information products designed by their organization/employers in order to determine if these products contain either culturally specific images or globally recognized images.

9.3.3 Cultural Briefing Assignment

Because all the graduate students in Mercer's MSTCO program are employed full time while they are taking classes, it is important to make the assignments relevant to their work. A work-related assignment both maintains student interest in and enthusiasm for the course, and it provides new and varied perspectives all involved parties can bring to the class discussions Accordingly, the goal of the first major class assignment—the cultural briefing assignment—is to help the students translate the cultural theories they have been examining into practical applications that will be useful when interacting with customers and coworkers.

To achieve this objective, the students form three-person teams. Each team then selects a culture (one other than their own native culture) for review, and each team then does research on that culture and prepares and delivers a 15–20 minute oral presentation, via WebEx, on strategies for communicating effectively with members of that culture. Assignment instructions are shown in Figure 9.1, while a grading rubric for the assignment is shown in Table 9.1.) Students also prepare briefs on cultures relevant to their work environments (e.g., cultures of their customers or coworkers).

> **Cultural Briefing Assignment**
>
> Using the resources provided in this class, as well as your own research, create an oral briefing for your co-workers. Your briefing should be designed to achieve the following:
>
> - Make your co-workers aware of cultural differences related to social and communication expectations.
> - Explain to co-workers how such differences might cause cross-cultural communication problems when they interact with individuals from these cultures.
> - Provide a list of strategies that co-workers can use to avoid potential mistakes based on this cultural information.
>
> The goal of this cultural profile is to provide guidance to your co-workers on how to effectively communicate with members of the target culture orally, visually, or in writing. Members of the target culture may consist of your current or potential customers or clients, co-workers, or vendors/suppliers. Limit your briefing to one culture.
>
> Briefing should last ~15–20 minutes.
>
> Use PPT to prepare the briefing. Use the PPT notes view to record the text of your briefing. Deliver via WebEX on the assigned date.
>
> Peer review process: Email the briefing to a classmate for peer review. Complete the review, using the rubric, and return to the authors within 72 hours.

FIGURE 9.1. Instructions for cultural briefing assignment. The purpose of this assignment is to apply cultural theories to workplace communication practices.

Such an exercise has proven quite successful, as indicated by student comments noting the value of this assignment:

- Student A: "I actually enjoyed the presentations and thought that they were not only an engaging way to learn about many different cultures in a shorter time span, but also a way to become more acquainted with our classmates, those in our groups and in other groups. The presentations allowed me to get a glimpse of each personality and allowed everyone to provide a more detailed interpretation of the models than what we had been able to do during class discussion. The presentations forced us to really gather our thoughts."
- Student B: "One very big advantage of the oral briefings is that everyone benefitted from the in-depth research done by one group. The class as a whole received specific information regarding 6 distinct cultures in a manner that is memorable enough to be a 'I remember something about that, I better look it up before we make a huge mistake' warning. This warning also helps us to make a viable initial argument for the time needed to investigate specifics regarding other cultures with whom we may interact."
- Student C: "Sharing ideas and discussing the cultural models with respect to a specific culture as a team, as well as peer-to-peer, expanded my appetite to discover more. The opportunity also opened the avenue to communicate through the use of graphics."

TABLE 9.1. Rubric for Evaluating Cultural Briefing Assignment. Students use the rubric to provide feedback to each other prior to delivering the final version of their presentation

Content	Strongly Agree	Agree	Neutral	Disagree	Strongly Disagree
Key cultural differences related to social and communications expectations are clearly explained.					
Analysis of cultural differences is based on one or more cultural models (Hofstede, Hall, Trompenaars, etc.)					
Potential problems caused by cross-cultural communication issues are identified.					
Strategies to communicate effectively with members of the target culture are clearly described.					
PPT notes view contains the script of the briefing and any other comments needed to understand the briefing content.					

Structure and Style	Strongly Agree	Agree	Neutral	Disagree	Strongly Disagree
The goal of the briefing is clearly articulated at the beginning of the document.					
The briefing been copyedited to ensure that it is free from grammatical and mechanical errors.					
Graphics reinforce content and are appropriately labeled.					
The style and tone are appropriate to the subject and audience.					
Content is appropriately chunked and organized.					
Overall evaluation	A	B	C	D	F

In summary, through these readings, discussions, and assignment, students who started this class with a limited awareness of the impact of culture on communication begin to develop an appreciation of the ways in which culture influences how people write and speak.

9.3.4 Feasibility Report Assignment

The second major class assignment, the feasibility report, is comprised of two main parts:

- Part 1: Each student chooses an information product designed by his or her organization
- Part 2: Students evaluate the options for making this product ready for distribution in a global market

The deliverable for this assignment is a feasibility report in which students examine the available options for tailoring a US information product (one designed by their employers) for use in a market outside of the United States. A description of this assignment and a copy of the rubric used to assess this assignment are shown in Figure 9.2 and Table 9.2.

Here are some examples of student responses to this assignment:

- Translate customer service documentation for Latin American affiliates of a large bank: identify in-house translation resources or third-party translation services
- Globalize the website of a large U.S. Department of Energy laboratory (by including international variations on country-specific information such as numbers, dates, times) or localize the website: create a second version of the web site that is specifically developed for another country by translating the content into another language and cultural context
- Localize product documentation to meet the Canadian requirements for French as well as English: outsource work to a Canadian software localization company or hire translators to work in-house
- Localize a mail-order catalog of women's clothing for Spanish-speaking customers: compare translation vendor services
- Globalize or localize time-and-attendance software for Central and South American markets: modify user interface to contain only icons or a mixture of icons and translated text, or completely translate the interface.
- Localize loyalty cards for banking customers in China: compare six visual designs for cards

The second time this course was taught, students asked if they could do more than a feasibility report; they actually wanted to prepare their information products to be translated because many of them anticipated having to do this in the future. After reviewing the literature on translation, the author eventually created an assignment (see Figure 9.3)

Feasibility Report Assignment

The CEO of your company and several high-level executives realize that to stay competitive, the company must begin pursuing some global initiatives. They have asked key players in the company, including you, to identify opportunities to provide more global products or services or to attract international clients.

Perform a feasibility study

You believe that a product or service that you provide can be globalized or localized. First, perform a feasibility study:

- Identify the opportunity
- Establish criteria for responding to the opportunity
- Determine the options
- Study each option according to the criteria
- Draw conclusions about each option
- Formulate recommendations

Write feasibility report

Write a 4–6 page feasibility report to your manager. Use the executive organization pattern.

- Introduction
- Factual summary
- Conclusions
- Recommendations
- Appendices

Give and receive one peer review

E-mail the profile to a classmate for peer review. Peer reviews must be completed and returned to the authors within 72 hours. Copy me on your completed PR.

Edit your report to incorporate any needed changes and revisions.

FIGURE 9.2. Feasibility report assignment. Students select an information product from their work setting and assess the feasibility of making the product ready for distribution in a global market.

based on the work of Maylath [47], an expert on translation issues and a contributor to this book. Table 9.3 illustrates the rubric for peer review and grading.

For this assignment, students select a brief (5–10 page) document designed by their organization, edit the document so it can be more easily translated, create a glossary of terms, and calculate the costs to translate their document into Spanish and Japanese. One of the most significant aspects of this assignment is the requirement that students describe exactly what changes they made to the document and why—something that forces them to provide the rationale behind their editing.

TABLE 9.2. Rubric for Evaluating Feasibility Report. Reports must evaluate several options for preparing information products for use in non-US markets

	Elements of Review—Feasibility Report
Audience and purpose	• Shows awareness of business audience • Written for the reader, not for the writer • Tells the writer's purpose in writing, focuses on action
Content	• Problem statement clear and well expressed • At least two potential solutions are presented • Body of report provides adequate support for conclusions • Options and criteria are clearly presented • Recommendation focused within organizational context of that business
Organization and order	• First paragraph contains all information required: subject, purpose, recommendations, overview • Background or Introduction is concise, complete, well organized for the target audience • Methods or Research explains options and criteria • Conclusions & Recommendations summarize findings and make a specific, business-focused recommendation • References (if needed) are presented in format the reader can use easily (APA style is fine, but may be less formal)
Format	• Memo format with all required information, including initials and job titles • Effective headings reveal the main divisions of report • 2–4 pages single spaced, double-spaced between paragraphs • Good use of tables, figures, or other visuals to enhance readability
Style and tone	• Written in the style of a business communication, not an academic paper • Uses lots of concrete words, good action verbs • Appropriate topic sentences for every paragraph, with rest of sentences supporting that thought • Polite and professional tone throughout
Grammar and punctuation	• Follows conventional standards for written English • No spelling errors (−2 points each) • No comma splices! (−5 points each) • Minimal grammatical errors (like punctuation) • Strong and clear sentence structure

Translation Assignment

The CEO of your company and several high-level executives have been asked by the marketing department to explore the feasibility of translating the company's documentation into Spanish and Japanese. The marketing department has received several requests from local customers for such documentation. Therefore, you have been asked to select a set of instructions and prepare them for translation as a test case to determine the feasibility and cost involved in translating company documents.

Prepare a set of instructions (or other documentation) for translation

Identify an existing set of instructions or other documentation in your organization. Next, make the text as clear, simple, consistent, and translatable as you can by following these guidelines:

- Identify words that may cause difficulty. Determine whether they are both precise and familiar. Define specialized terms in an accompanying glossary.
- Identify and eliminate idioms by substituting words with straightforward meanings.
- Make sure all terms are used consistently in meaning throughout the text.
- Eliminate humor (which is often untranslatable).
- Edit sentences to make them relatively short independent clauses.
- Leave room on the page for simultaneous translation or at the end of the instructions for consecutive translation, where other languages will appear. Many languages take 30% more space than does the English language to say the same message.
- Find out as much as you can about cultural and rhetorical differences and adapt the text. For instance, Americans like a summary introduction, but Japanese readers often want a long introduction of theoretical principles before getting to the steps involved.

Deliverables for assignment

- A set of instructions or other documentation that have been prepared for translation according to the guidelines above. Include the original version for comparison.
- A glossary defining specialized terms that appear in the instruction set
- A cost estimate for translating the instructions into Spanish and Japanese.
- A brief description of how you prepared the document to be translated.

Give and receive one peer review

E-mail the profile to a classmate for peer review. Peer reviews must be completed and returned to the authors within 72 hours. Copy me on your completed PR.

Edit your materials to incorporate any needed changes and revisions.

FIGURE 9.3. Translation assignment. Students select an information product from their work setting and prepare it for translation into Spanish and Japanese. *Source:* Adapted from Reference 47 with permission.

Resources

- Kohl J. 1999. Improving translatability and readability with syntactic cues. TC, 46(2), 149-166.
- Kohl J. 2008. *The Global English Style Guide*, SAS.
- *Translation, Localisation, Interpreting and Website Translation* articles—great collection of resources (http://www.kwintessential.co.uk/translation/articles.html)
- *Global Advisor newsletter*—some excellent tips on writing for translation (http://www.intersolinc.com/newsletters/newsletter_23.htm)
- *Writing for Translation—Guide to Getting Started*—series of articles and links to translation vendors; worth reading (http://www.multilingual.com/downloads/screenSupp107.pdf)
- *Technical Writing for Translation*—GITS—interesting comments about preparing documentation for Japanese clients (http://ginstrom.com/translation/tech_writing.php)
- *Translation Directory*—portal for freelance translators (http://www.translationdirectory.com/article427.htm)

FIGURE 9.3. *(Continued).*

TABLE 9.3. Rubric for Evaluating Translation Assignment. The review process enables students to evaluate translation issues in different genres of information products

	Elements of Review—Translation Report
Audience and purpose	• Shows awareness of business audience • Written for the reader, not for the writer • Tells the writer's purpose in writing
Overview	• Describes document that was chosen to be translated and rationale for its selection • Describes what was done to prepare document for translation (style choices, word choices, terms, editing for sentence length, etc.) • Describes cultural considerations for translating
Cost estimate	• Includes cost estimates to have document translated into Spanish and Japanese • Describes how costs were calculated
Glossary	• Includes specialized terms in the document • Includes part of speech for each term • Defines each term defined and gives an example of how it is used in a sentence
Documents	• Includes original document • Includes document prepared for translation • Follows guidelines for editing text so that it is clear, simple, and easily translatable (use the editing checklist as a guide; see homepage for link)

Excerpts from previous student reports reveal the types of documents students can select for this assignment as well as the approaches they can use to complete this assignment:

- "My organization did not plan for translation. We took many liberties with the terminology, syntax, and grammar within our user guides. In preparing the content for translation, I created a glossary, edited the content, modified the style, and considered how the graphics should be altered and what role source and target culture would play." [Document was a user guide for customer order management.]
- "I rewrote the content to avoid noun clusters, present participles (-ing verb form), passive voice (within reason), long sentences, and ambiguous terms. I also removed the i symbol from Note paragraphs. The lowercase "i" has no meaning in Japanese." [Document was a user guide.]
- "The document was analyzed specifically for uses of humor, metaphors, similes, modifier strings, adverbs representing time (when, where, there, while), relative pronouns (who, which, that, whose), gerunds, acronyms, redundancy, symbols, and long sentence structures. Each sentence was re-written in simple sentence structures and with independent clauses." [Document was an IT "how to" manual.]
- "Because the workflow flowchart must be included with each workflow document (per management decision), a .tiff version of this flowchart will be provided to the translation company for their use in creating a translated version. Plenty of space is available on the page for an expanded version." [Document was a guide for software workflow.]

In summary, this portion of the course introduces students to the issues involved in localizing information products for culturally diverse customers and allows them to put theory into practice.

9.4 Conclusion

This chapter has provided an overview of the rationale behind a course in international technical communication in the Master's in Technical Communication Management at Mercer University. In so doing, the chapter has also provided an overview of the content of the course and course assignments. The author has taught this course twice as an elective, and due to the success of the course—and the recognition among students (and employers) of the need for such a course—it is now included as a required element in the curriculum due to the need to educate graduate students to be ready to participate in the global economy.

The course covers theories on culture, cultural models, rhetorical patterns, internationalization and localization issues, translation, and graphic design for international audiences. It also gives the students an opportunity to learn about and then implement best practices in preparing information products for international markets: practices which are always constrained by cost, schedule, time, and personnel resources. In this way, the class reflects the realities of the modern, globalized workplace.

In sum, individuals who wish to participate successfully in the global economy must understand how to communicate across cultural, national, and economic borders. Many academic institutions have recognized this need and have added courses to prepare their students to participate in this flattened world. However, it is imperative that educators in technical and professional communication reach out to institutions and businesses outside the United States in order to share best practices as globalization progresses.

References

1. G. Hofstede and G. J. Hofstede, *Cultures and Organizations: Software of the Mind*. Maidenhead, England: McGrawHill, 2005.
2. B. McSweeney, "Hofstede's model of national cultural differences and the consequences: A triumph of faith - a failure of analysis," *Human Relations*, vol. 55, no. 1, pp. 89–118, January 2002.
3. P. Signorini, R. Wiesemes, and R. Murphy, "Developing alternative frameworks for exploring intercultural learning: A critique of Hofstede's cultural difference model," *Teaching in Higher Education*, vol. 14, no. 3, pp. 253–264, June 2009.
4. E. T. Hall, *Beyond Culture*. New York: Random House, 1976
5. P. W. Cardon, "A critique of Hall's contexting model: A meta-analysis of literature on intercultural business and technical communication," *Journal of Business and Technical Communication*, vol. 22, no. 4, pp. 399–424, September 2008.
6. E. Gould, "Synthesizing the literature on cultural values," in *Usability and Internationalization of Information Technology*, N. Aykin, Ed. Mahwah, NJ: Erlbaum, 2005, pp. 79–121.
7. F. M. Zahedi, W. V. VanPelt, and J. Song, "A conceptual framework for international web design," *IEEE Transactions on Professional Communication*, vol. 44, no. 2, pp. 83–103, June 2001.
8. E. Callahan, "Cultural similarities and differences in the design of university websites," *Journal of Computer-Mediated Comunication*, vol. 11, no. 1, article 12, 2005. DOI:10.1111/j.1083-6101.2006.tb00312.x
9. J. C. Usunier and N. Roulin, "The influence of high- and low-context communication styles on design, content, and language of business-to-business web sites," *Journal of Business Communication*, vol. 47, no. 2, pp. 189–227, April 2010.
10. L. Rosado, "Cross-cultural communications: A Latino perspective," *AE Extra*, 2005. Available at: http://www.unco.edu/ae-extra/2005/1/Art-1.html (accessed August 2, 2015).
11. B. Thatcher, "Cultural and rhetorical adaptations for South American audiences," *Technical Communication*, vol. 46, no. 2, pp. 177–195, May 1999.
12. W. Coggin, and B. F. Coggin, "So you want to work in China," *Technical Communication*, vol. 48, no. 4, pp. 389–396, November 2001.
13. R. Schaler, "Communication as a key to global business," *Proceedings of the 2005 IEEE International Professional Communication Conference*, 2005, pp. 1–10.
14. D. Walmer, "One company's efforts to improve translation and localization," *Technical Communication*, vol. 46, no. 2, pp. 230–237, May 1999.

15. H. Yli-Jokipii, "The local and the global: An exploration into the Finnish and English websites of a Finnish company," *IEEE Transactions on Professional Communication,* vol. 44, no. 2, pp. 104–113, June 2001.
16. F. Sapienza, "Nurturing translocal communication: Russian immigrants on the world wide web," *Technical Communication*, vol. 48, no. 4, pp. 435–448, November 2001.
17. H. Sun, "The triumph of users. Achieving cultural usability goals with user localization," *Technical Communication Quarterly*, vol. 15, no. 5, pp. 457–481, Autumn 2006.
18. N. St. Germaine-Madison, "Instructions, visuals, and the English-speaking bias in technical communication," *Technical Communication*, vol. 53, no. 2, pp. 184–194, May 2006.
19. J. H. Melton, "Lost in translation: Professional communication competencies in global training contexts," *IEEE Transactions on Professional Communication,* vol. 51, no. 2, pp. 198–214, June 2008.
20. W. B. Potsus, "Is your documentation translation-ready?," *Intercom*, vol. 48, no. 5, pp. 12–17, May 2001.
21. P. Flint, M. L. Van Slyke, D. Starke-Meyerring, and A. Thompson, "Helping technical communicators help translators," *Technical Communication*, vol. 53, no. 2, pp. 238–248, November 1999.
22. B. Sichel, "Planning and writing for translation," Writing for Translation Guide: Getting Started, October/November 2009. Available at: http://www.multilingual.com/downloads/screenSupp107.pdf (accessed August 2, 2015).
23. B. Maylath, "Writing globally: Teach the technical writing students to prepare documents for translation," *Journal of Business and Technical Communication*, vol. 11, no. 3, pp. 339–352, July 1997.
24. M. Flammia, "Preparing technical communication students to play a role on the translation team," *IEEE Transactions on Professional Communication,* vol. 48, no. 4, pp. 401–412, December 2005.
25. N. R. Goodman, "Cross-cultural training for the global executive," in *Improving Intercultural Interactions: Modules for Cross-cultural Training Programs*, R. W. Brislin and T. Yoshida, Eds. Thousand Oaks, CA: Sage Publications, 1994.
26. N. Hoft, "Global issues, local concerns," *Technical Communication*, vol. 46, no. 2, pp. 145–148, May 1999.
27. G. Hayhoe, "The globalization of our profession," *Technical Communication*, vol. 53, no. 1, pp. 9–10, February 2006.
28. N. Hoft, *International Technical Communication: How to Export Information about High Technology*. NY: John Wiley & Sons, 1995.
29. P. Honold, "Learning how to use a cellular phone: Comparison between German and Chinese users," *Technical Communication*, vol. 46, no. 2, pp. 196-205, May 1999.
30. A. Marcus, "User interface design and culture," in *Usability and Internationalization of Information Technology*, N. Aykin, Ed. Mahwah, NJ: Erlbaum, 2005, pp. 51–78.
31. U. Connor, "New directions in contrastive rhetoric" *TESOL Quarterly*, vol. 36, no. 4, pp. 493–510, 2002.
32. C. Barnum and H. Li, "Chinese and American technical communication: A cross-cultural comparison of differences," *Technical Communication*, vol. 53, no. 2, pp. 143–166, May 2006.
33. M. Gerritsen and E. Wannet, "Cultural differences in the appreciation of introductions of presentations," *Technical Communication*, vol. 52, no. 2, pp. 194–208, May 2005.

34. J. H. Spyridakis and W. Fukuoka, "The effect of inductively versus deductively organized text on American and Japanese readers," *IEEE Transactions on Professional Communication,* vol. 45, no. 2, pp. 89–114, June 2002.
35. Y. Wang and D. Wang, "Cultural contexts in technical communication: A study of Chinese and German automobile literature," *Technical Communication*, vol. 56, no. 1, pp. 39–50, February 2009.
36. C. Campbell, "Beyond language: Cultural predispositions in business correspondences," presented at the Region 5 Society for Technical Communication Conference, Fort Worth, TX, 1998.
37. J. Wang, "Toward a critical perspective of culture: Contrast or compare rhetorics," *Journal of Technical Writing and Communication*, vol. 38, no. 2, pp. 133–148, 2008.
38. N. Aykin, "Overview: Where to start and what to consider," in *Usability and Internationalization of Information Technology*, N. Aykin, Ed. Mahwah, NJ: Erlbaum, 2005, pp. 3–20.
39. M. McCool, "Information architecture: Intercultural human factors," *Technical Communication*, vol. 55, no. 2, pp. 167–183, May 2006.
40. R. de Pedro, "The translatability of texts: A historical overview" *Meta*, vol. 44, no. 4, pp. 547–559, 1999.
41. J. House, "Translation quality assessment: Linguistic description versus social evaluation," *Meta*, vol. 46, no. 2, pp. 243–257, 2001.
42. T. Altanero, "Translation MT and TM," *Intercom*, vol. 47, no. 5, pp. 24–26, May 2000.
43. J. W. Akis and W. R. Sisson, "Improving translatability: A case study at Sun Microsystems, Inc.," *The Globalization Insider,* 2006. Available at: http://archive.is/BpOZD (accessed August 2, 2015).
44. W. Horton, "Graphics: The not quite universal language," in *Usability and Internationalization of Information Technology*, N. Aykin, Ed. Mahwah, NJ: Erlbaum, 2005, pp. 157–187.
45. B. Moore, "Designing for multicultural and international audiences. Creating culturally intelligent visual rhetoric and overcoming enthnocentrism," MA thesis, Department of English, University of Central Florida, Orlando, FL, 2010.
46. Q. Wang, "A cross-cultural comparison of the use of graphics in scientific communication," *Technical Communication*, vol. 47, no. 4, pp. 553–560, November 2000.
47. B. Maylath, "Translating user manuals: A surgical company's 'quick cut'," in *Global Contexts. Case Studies in International Technical Communication*, D. Bosley, Ed. Boston, MA: Allyn & Bacon, 2001, pp. 64–80.

10

Teaching Cultural Heuristics Through Narratives: A Transdisciplinary Approach

Han Yu

Kansas State University

This chapter argues for the instructional value of cultural heuristics (or models, tools, and frameworks, as they are variously called) and describes a pedagogical approach that teaches heuristics through cultural narratives. This pedagogy combines the functionalist and interpretive approaches to understanding culture and offers a transdisciplinary method that is constructive in teaching intercultural communication and relevant to engineering students and professionals. In examining these ideas, the chapter overviews commonly referenced cultural heuristics and introduces four types of cultural narratives. The chapter also describes how teachers can use these narratives to teach heuristics and extracultural knowledge in context.

10.1 A Transdisciplinary Approach for Global Engineers

For over a decade, engineering programs in the United States have recognized the importance of preparing students to be global engineers who possess, among other qualities, the ability to communicate with colleagues and clients across regions and cultures [1]. In this chapter, I discuss a pedagogical approach that can aid engineering students' global

---

Teaching and Training for Global Engineering: Perspectives on Culture and Professional Communication Practices, First Edition. Edited by Kirk St.Amant and Madelyn Flammia.
© 2016 The Institute of Electrical and Electronics Engineers, Inc. Published 2016 by John Wiley & Sons, Inc.

education and prepare them for the kind of intercultural interactions they are likely to encounter in the workplace. This transdisciplinary pedagogy combines the functionalist and interpretative paradigms to teach cultural heuristics through narratives. By doing so, this approach can offer cultural instruction that is engaging, effective, and relevant to engineering.

> A transdisciplinary pedagogy combines the functionalist and interpretive paradigms to teach cultural heuristics through narratives.

In proposing this pedagogy, I'm aware of the critiques leveled at cultural heuristics, both within and beyond the field of technical and professional communication. It is my argument, however, that teachers should be careful about how they interpret these critiques. Incomplete or mistaken interpretations of the critiques might cause teachers and students to reject valuable tools that can be used to study intercultural communication. We can avoid such hasty rejections by understanding the instructional values of cultural heuristics and by teaching these heuristics within contextual narratives of culture.

In this chapter, I introduce some commonly referenced cultural heuristics to establish a point of reference. I then discuss common critiques of these heuristics as well as counter-arguments that challenge the critiques. In so doing, I argue that much of this scholarship debate is a result of paradigmatic differences in research. I also argue that it is more productive if we supplement, rather than replace, cultural heuristics in our teaching. I then review several types of cultural narratives and explain how they can be used as a platform to teach heuristics. Finally, limitations of this pedagogy are discussed.

10.2 Overview of Cultural Heuristics

Cultural heuristics are theoretical frameworks that examine and compare how members of different cultures orient themselves toward certain values or dimensions believed to be universally relevant (i.e., as applying to most or all cultures). Many cultural heuristics have been developed by researchers, teachers, and cultural trainers. Here, I review some well-known and commonly referenced heuristics in technical and professional communication: Hall's concepts of time, space, and context; Hofstede's Five-Dimension Heuristic, and Trompenaars and Hampden-Turner's Seven-Dimension Heuristic. Critiques, counter-critiques, and applications of these heuristics are discussed in following sections.

10.2.1 Edward T. Hall's Concepts of Time, Space, and Context

Edward T. Hall was an American anthropologist and cross-cultural researcher whose work has had significant influence on the study of intercultural communication. His work emphasizes "the importance of nonverbal signals and modes of awareness over explicit messages" [2]. One of Hall's most influential concepts involves perceptions of time and space. Hall theorized two ways that cultures approach time: the *monochronic time (M-time)* and the *polychromic time (P-time)*. In discussing M-time and P-time, Hall wrote that they represent two variant solutions to the use of both time and space as organizing frames for activities.... M-time emphasizes schedules, segmentation, and promptness. P-time systems are characterized by several things happening at once. They stress involvement of people and completion of transactions rather than adherence to preset schedules [3, p. 14].

Another concept that is essential to Hall's cultural heuristic is context. Hall theorized that cultures approach context on a continuum of high-context (HC) and low-context (LC) communication. As he explains,

A high-context (HC) communication or message is one in which most of the information is either in the physical context or internalized in the person, while very little is in the coded, explicit, transmitted part of the message. A low-context (LC) communication is just the opposite; i.e., the mass of the information is vested in the explicit code [3, p. 79].

To someone who is not used to HC communication, this approach might seem indirect, even mysterious, because all things are not explicitly spelled out in the transmitted message. On the other hand, to someone who is not used to LC communication, this communication style might seem blunt or even rude, for all messages, including negative ones, are directly conveyed.

10.2.2 Geert Hofstede's Five-Dimension Heuristic

Geert Hofstede is a Dutch social psychologist and anthropologist. Based on his survey studies of 116,000 IBM employees in 72 countries in the 60s and 70s, Hofstede [4] proposed a cultural heuristic that contains five dimensions:

- *Power distance*, which describes the extent to which members of a culture "accept and expect that power is distributed unequally" [4, p. xix]
- *Uncertainty avoidance*, which describes the extent to which members of a culture feel comfortable in unknown or unstructured situations
- *Individualism versus collectivism*, which describes the extent to which members of a culture identify with self or group
- *Masculinity versus femininity*, which describes the extent to which members of a culture exhibit masculine (such as assertive and achievement-oriented) or feminine (such as nurturing and relationship-oriented) thinking patterns and behaviors
- *Long-term versus short-term orientation*, or the *Confucian dynamism*, which describes the extent to which members of a culture "accept delayed gratification of their material, social, and emotional needs" [4, p. xx].

According to Hofstede, these five dimensions can be used to understand and even predict how humans in different value systems think, feel, and behave as well as how their organizations and institutions operate [4]. Hofstede's heuristic has generated numerous replication studies and discussions since its publication in 1980, and his book *Culture's Consequences*, which first introduced this heuristic, remains one of the most cited sources in the *Social Science Citation Index* [5].

10.2.3 Fons Trompenaars and Charles Hampden-Turner's Seven-Dimension Heuristic

Fons Trompenaars is a Dutch cross-cultural communication consultant, and Charles Hampden-Turner is a British business and management theorist. Based on their experience conducting cross-cultural training programs in over 20 countries and drawing on the responses of 30,000 research participants from 30 multinational companies, Trompenaars and Hampden-Turner introduced a cultural heuristic that consists of seven dimensions in their book *Riding the Weaves of Culture* [6]:

- *Universalism versus particularism,* which examines the extent to which members of a culture value universal, abstract rules or particular, contextual circumstances
- *Individualism versus communitarianism,* which is similar to Hofstede's *individualism-collectivism* dimension
- *Neutral versus affective,* which examines the extent to which members of a culture are comfortable with objective, detached interactions or with feelings and emotions
- *Specific versus diffuse,* which describes the extent to which members of a culture separate personal and professional lives and relationships
- *Achievement versus ascription,* which describes the extent to which members of a culture confer status based on one's recent achievements or personal background
- *Attitudes to time,* which describes the extent to which members of a culture value the past, the present, or the future
- *Attitudes to the environment,* which describes the extent to which members of a culture believe they should control nature or go along with nature

According to Trompenaars and Hampden-Turner, these dimensions explain the unique solutions different cultures use in order to cope with universal problems: how to deal with fellow human beings, how to deal with time, and how to deal with environment [6, p. 8]. This heuristic is particularly popular within the field of international business and management, and Trompennars, for his work in and beyond *Riding the Waves of Culture*, is considered one of today's most influential management thinkers.

10.3 Critiques and Counter-Critiques of Cultural Heuristics: How to Move Forward from Misguided Debates

The cultural heuristics reviewed in the previous section are widely cited and have significant influence in a wide range of disciplines such as business, management, organizational psychology, and communication—including intercultural technical and professional communication. At the same time, these heuristics also invited much criticism, both within and beyond the field of technical and professional communication. Hall's concept of context, for example, is critiqued for lacking research rigor, empirical evidence, measurement instruments, and subsequent validation [7]. More generally, because Hall's work is largely based on his "actual encounters with the cultural Other," it is said to be "hasty generalizations" [8]. Hofstede's study attracted more critiques. Some argue that he overlooked the "fuzziness" of culture and contextual elements and treated national culture as a static concept [9]. Other critics argue that certain dimensions in his heuristics (e.g., individualism and Confucian dynamism) are interrelated and might not be conceptually separate [10]. Still others focus on the limitations of particular dimensions. Fang, for instance, argues that Hofstede's fifth dimension is derived from redundant value options in his survey studies and mistakenly treats dualistic Chinese values as oppositional ones [5].

These questions and critiques are valuable, for they facilitate the continuous refinement of cultural heuristics and caution readers about the complexity of the heuristics and the difficulty of understanding and using them properly. However, it becomes dangerous when these critiques present, sometimes unintentionally, the impression that heuristics are a doomed approach and should be rejected altogether in the teaching of culture. This impression has become a realistic danger in technical and professional communication as the field's study of intercultural issues grew increasingly sophisticated in recent years and as teachers and scholars became more reflective in their work. But all of these critiques, as I argue in the next section, cannot be simplistically accepted (or rejected, for that matter). Instead, more inclusive grounds can and should be sought to combine different understandings of heuristics and to advance the pedagogies of intercultural technical and professional communication.

10.3.1 Misattributed and Unsystematic Critiques

In "Culture and cultural identity in intercultural technical communication," Hunsinger [11] gives a convincing critique of the field's use of heuristics. To be precise, Hunsinger does not evaluate the heuristics themselves (e.g., their construct, measurement, or validation). Rather, he examines the pedagogical approach of using heuristics to teach intercultural technical communication. Even though Hunsinger concludes that we should supplement rather than replace heuristics in our teaching, the strong argument made in his article can lead some readers to conclude that heuristics should be rejected altogether.

Hunsinger draws upon important figures in technical and professional communication to support his critique. He, for example, quotes Edmond Weiss to argue that the heuristic approach is "an essentializing practice" that "'treats members of a group as instances of a profile'" [11, p. 33]. A reading of Weiss [12], however, shows that he does not really focus on the teaching of cultural heuristics. Rather, Weiss is concerned that uncritical teaching of international communication (whether it is cultural heuristics or other contents) might be reduced to the mercury goal of promoting business in foreign countries without the consideration of ethics, honesty, and justice. Furthermore, when Weiss cautions that we should not treat members of a group as instances of a profile, he is critiquing the teaching of "facile generalizations" such as "blue represents truth in Arabic culture but villainy in Japanese," "red signals danger in Europe or Japan but festivity in China," or "an American businesswoman should not offer to shake hands with an Egyptian businessman" [12, p. 260]. These kinds of generalizations are *not* what cultural heuristics are about. Indeed, heuristics attempt to provide the deeper, more complex understandings about culture that Weiss calls for.

In his critique, Hunsinger [11] also cites Beamer's article "Finding a way to teach cultural dimensions" [13] to argue that the heuristic approach might be "based on limited research or unrepresentative ethnographic data" [11, p. 33]. It is true that, in her article, Beamer does voice her doubt that the IBM employees in Hofstede's study "totally represented the general culture" [13, p. 112]. Yet she does not offer evidence to prove her doubt. Indeed, in her article, Beamer seems to accept experience-based cultural dimensions and is not concerned with statistical representativeness or mathematical measurement. In the same article, Beamer offers her own eight-dimension heuristics without explaining how these dimensions came about and states that many authors, teachers and trainers "come up with lists of cultural values that make it easier to compare cultures" [13, p. 115]. Beamer also admits that the visualization technique she developed to teach her heuristic "has no basis in statistical or mathematical measurement" but is a "gut call" [13, p. 117].

Finally, Hunsinger cites Munshi and McKie [14] to argue that the heuristic approach can "misrepresent cultural identity" and emphasize "differences that matter" [11, p. 33]. Munshi and McKie do critique how many intercultural business communication textbooks draw upon Hofstede to generalize and categorize national cultures. This categorization, Munshi and McKie argue, creates a "binary division between the largely Western "we" and the largely non-Western "they"" and promotes the neo-colonial desire to master the less economically advanced cultural other [14, p. 14]. This critique, however, can be best understood in terms of different research paradigms, which I discuss in the following section.

10.3.2 Differences in Research Paradigms

In his critique of Hofstede, subtitled "a triumph of faith – a failure of analysis," McSweeney [15] argues that Hofstede's heuristic rests on five assumptions. McSweeney goes on to claim that each assumption is false, and therefore, the entire heuristic must be rejected. For instance, McSweeney argues that Hofstede had incorrectly assumed that national cultures can be measured at the individual level and can be represented by IBM employees who must have "many nationally atypical characteristics" [15, p. 101].

Williamson [16], in turn, critiques McSweeney by noting that each of McSweeney's arguments is, in itself, flawed. The point about IBM employees, for instance, "confuses the phenomenon of culture with its measurement" [16, p. 1380]. Because culture is an abstract construct for which there is no direct measure, no matter which sample populations are studied (for instance, teachers or students), they can likewise be criticized for lacking representativeness. Furthermore, "Hofstede's dimensions are not absolute measures, but relative positions by which nations can be compared....[T]he issue is not whether the samples are representative of national populations, but whether differences between their responses are representative of differences in cultural values" [16, pp. 1380–1381].

Williamson's most insightful observation, however, is that McSweeney adopts two incompatible research paradigms in his critique of Hofstede: the functionalist and the interpretive. As Burrell and Morgan [17] explain, social science research can be classified into either a functionalist paradigm or interpretive paradigm. The distinction depends on whether the researcher takes an objective or subjective approach to ontology (the perception of the world), epistemology (the perception of knowledge), human nature, and research methodology.

The functionalist paradigm sees the world as external and "real"; assumes positivist epistemologies and sees humans as being controlled by the world. This paradigm also adopts a nomothetic research methodology that emphasizes systematic protocols, scientific tests, quantitative techniques, and standardized research instruments such as surveys, questionnaires, and personality tests [17]. This research paradigm "tends to assume that the social world is composed of relatively concrete empirical artifacts and relationships which can be identified, studied, and measured through approaches derived from the natural sciences" [17, p. 26]. The search for "universal laws" is a most forceful expression of the functionalist paradigm's concern [17, p. 3].

Heuristic scholars such as Hofstede generally operate from the functionalist paradigm. They assume that national culture is "a given regularity that shapes shared values" and attempt to study it using "statistical techniques designed to suppress subjective interpretations" [16, p. 1375]. They also hope to design models that "purport to be universally applicable" [16, p. 1375].

The interpretive paradigm, by contrast, sees the world as internal and constructed. It assumes anti-positivist epistemologies and sees humans as controlling their environment. It also favors the ideographic methodology, which emphasizes the analysis of subjective accounts one generates by getting close to one's subjects and getting inside their life situations [17]. This research paradigm is concerned with understanding "the social world at the level of subjective experience. It seeks explanation within the realm of individual consciousness and subjectivity...[and] a network of assumptions and intersubjectively shared meanings" [17, pp. 28–29]. Rather than seeking universal laws, the interpretive paradigm seeks to understand "the subjectively created social world 'as it is' in terms of an ongoing process" [17, p. 31].

Hall's work, in turn, is largely based on the interpretive paradigm. It favors personal experiences, not clinically controlled empirical data. It does not isolate culture into exclusive dimensions for analysis, and it focuses on selective cultures rather than comprehensive models.

Given these different paradigms, attacking scholarship about culture, as Burrell and Morgan [17] would say, will be easy. Individuals can simply critique work located in one paradigm by positioning their assumptions and priorities in the other [17, p. 395]. McSweeney's [15] interpretive critique of Hofstede follows this strategy (although he also provides critiques from within the functionalist paradigm). "McSweeney appears to doubt that national cultures can be measured explicitly," an idea that "would sit uncomfortably with positivist epistemology" [16, p. 1390]. Instead, McSweeney advocates that researchers "engage with and use theories of action which can cope with change, power, variety, multiple influences—including the non-national—and the complexity and situational variability of the individual subject" [15, p. 113]. This call, Williamson writes, is well founded but "very difficult to meet with research using nomothetic protocols for carefully controlled variables" [16, p. 1390].

> Sitting uneasily between conflicting research paradigms is an important reason why technical and professional communication teachers have not been able to reach an agreement on whether and how to productively teach cultural heuristics.

Technical and professional communication scholars occupy an interesting place in between these two paradigms. On one hand, our historical connection with sciences and engineering pushes us toward the functionalist approach. At the same time, our training in rhetoric and the 70s backlash against positivism pushes us toward the interpretive approach. So, on one hand, we have critiques such as Cardon's [7], which operate from the functionalist paradigm and emphasize empirical evidence and measurement validation. On the other hand (and more frequently), we have critiques such as Hunsinger's [11] and Munshi and Mckie's [14], which embrace the interpretive paradigm and object to heuristics' seeming binary representations of culture and disregard of contextual elements. Sitting uneasily between conflicting research paradigms is an important reason why technical and professional communication teachers have not been able to reach an agreement on whether and how to productively teach cultural heuristics. And resolving this conflict, as I discuss in the following section, can offer the field a more productive pedagogical approach to teach culture.

10.3.3 Forward from the Critiques: A Transdisciplinary Approach to Integrate the Interpretive and the Functionalist Approaches

It is not my purpose to critique Hunsinger [11] or other scholars who object to cultural heuristics. Their works serve as important reminders that cultural heuristics are not without problems, that scholars should question or attempt to refine these heuristics, and that educators should be careful in how they teach them. At the same time, these critiques should be based on more comprehensive understandings of cultural heuristics and more serious attempts to engage with them. Our scholars should recognize that their research paradigms and conceptions of culture may be very different from authors such as Hall, Hofstede, or other heuristic scholars. Without this recognition, we will continue to mistake cultural patterns that are expressed in "bell-curve statistics" as "monolithic traits" [18, p. 7].

Recognizing this difference is the first step toward more accurate understandings and productive teaching of cultural heuristics. After that, we can then attempt to combine the functionalist and interpretive approaches to approach culture. As Williamson [16] writes, individuals are influenced by as well as influence their social worlds, and we need multiple approaches from multiple paradigms in the study of culture. Each paradigm offers pedagogical advantages. The interpretive approach helps students recognize complex and interconnected cultural and extracultural elements as they are operationalized in unique and changing contexts. The functionalist approach helps students identify a finite number of variables that have potential to explain multicultural varieties and promise some parsimoniousness in the unstructured social world. By combining the functionalist and interpretive approaches, technical and professional communication teachers can create a transdisciplinary approach that incorporates the best of both worlds, an approach that can lead to more comprehensive and effective teaching of culture.

This transdisciplinary pedagogy will be particularly relevant to engineering students and professionals because it reflects two dimensions of the engineering epistemology that co-exist in tension: the dimension of basic sciences and the dimension of design [19, p. 94]. The dimension of basic sciences, which is functionalist in nature, values analytical thinking and "views engineering as the application of the natural and exact sciences, stressing the values of logics and rigor, and seeing knowledge as designed through analysis and experimentation. Research is the preferred *modus operandi* of this dimension, where the discovery of first principles is seen as the activity leading to higher recognition" [19, p. 94]. Given this functionalist dimension in engineering curricula, students are likely to accept similar assumptions in the study of culture.

By contrast, the dimension of design, which is interpretive, values systems thinking, including "exploring alternatives and compromising" [19, p. 94]. This dimension "resorts frequently to non-scientific forms of thinking, [and] the key decisions are often based on incomplete knowledge and intuition, as well as on personal and collective experiences" [19, p. 94]. Engineering's recognition of this dimension provides an opportune ground for the interpretive teaching of culture with its multiple, interconnected, and shifting cultural and extracultural variables.

The question, then, becomes how to teach engineering students cultural heuristics in ways that emphasize both the functionalist and interpretive approaches. In the following section, I suggest one way to do so: teaching heuristics through narratives.

10.4 Overview of Cultural Narratives

According to scholars in anthropology, psychology, and sciences, narratives are a primary thought and discourse process that human beings use to form identities, understand experiences, develop rationality, build knowledge, and handle human interactions [20–22]. If narratives are such a fundamental part of our life, they should also be an important way for students to understand culture—their own cultures as well as other cultures. Several types of cultural narratives are available for this purpose: case studies, cross-cultural dialogs, critical incidents, and open stories.

10.4.1 Case Studies

Case studies describe problems that occur among certain characters and provide "enough detail to assess the problems involved and determine possible solutions" [23, p. 57]. They provide students with realistic views of the professional fields they are trained for or plausible scenarios they may encounter in those fields. Discussion questions or activities often accompany a case, and these items can help students collect additional information or consider and solve the problems presented in the case.

Case studies have been used for decades as an instructional device in business, law, medicine, engineering, and anthropology [23, 24]. Their use in intercultural studies started in the early days of the Peace Corps [23, p. 57]. Case studies are also used in teaching intercultural technical communication: comprehensive textbooks such as Markel [25] use them as one instructional method, and Bosley [26], in particular, offers a collection of case studies that address a range of issues and contexts in international technical communication.

Because case studies portray real-world situations and concrete problems "typical of those encountered in cross-cultural or ethnically diverse situations," they can be effective at engaging students [23, p. 57]. When students become involved in a case, they will actively consider options and assess strategies, thereby realizing intercultural learning [23].

10.4.2 Cross-Cultural Dialogs

Cross-cultural dialogs "are brief conversational excerpts between two people of different cultures, used to increase cultural self-awareness and to dramatize subtle differences between cultures" [23, p. 70]. In Storti [27, 28], a dialog contains 5–8 utterances, no background information precedes the dialog, and the only clue is the speakers' names which are suggestive of their nationalities. These cross-cultural dialogs can be grouped by settings (e.g., workplaces and schools) or intercultural interactions (e.g., between Americans and the British). Analyses of the dialogs are provided along with the text of the actual exchange and generally appear in a separate section to encourage students' independent learning.

Well-written cross-cultural dialogs are an interesting paradox. On one hand, they focus on the speakers' "different values, attitudes, or views of the world" [23, p. 70]. As Storti writes, it is not that people of different cultures have no similarities, but "similarities don't usually lead to misunderstandings" so "they are not food for dialogues" [28, p. 9]. On the other hand, the dialogs are very subtle about these differences and even the misunderstandings they cause [28]. This "not only makes the dialogue a puzzle which the reader is then challenged to figure out, but the invisible nature of the mistake also accurately reflects real-life cross-cultural encounters, where the players are often not aware that a cultural difference has caused a misunderstanding" [28, p. 6].

Cross-cultural dialogues also illustrate the comparative nature of cultural differences. While a dialogue between, say, an American and a British individual might illustrate the directness in American communication patterns, the next dialog between an American and a German individual will find that American communication is not too direct after all.

10.4.3 Critical Incidents

Critical incidents are short vignettes "in which individuals from different cultures interact with the intent of pursuing some common goal. Toward the end of each incident a clash of cultures is evident and the two parties are unable to accomplish their task" [29, p. 13]. Critical incidents differ from cross-cultural dialogs by using not just dialogs, but other narrative tools including characters, scenes, and context. In this sense, critical incidents are similar to case studies, although they are much briefer. In Cushner and Brislin, an incident is about 200–300 words, whereas in Bosley [26], a case study is several-thousand words.

Another distinct feature of critical incidents is that each incident is followed by several alternative explanations. A reader is then asked to select one or sometimes more that best account for the problem *"from the point of view of the actor in the vignette who is not a member of the reader's own culture"* [29, p. 13]. Doing so helps readers better understand unfamiliar cultures and make more accurate interpretations and attributions of others' behaviors [29]. Students might also write open-ended answers for the incidents, although Cushner and Brislin caution that doing so might cause subsequent discussions to lack focus and direction [29, p. 43]. Answers and explanations of the choices are provided, often in a separate section to encourage students' independent learning.

Critical incidents help students to become "more knowledgeable about concepts relevant to cross-cultural interactions," "to generate, analyze, and apply such concepts to personal cross-cultural incidents," and to become "more sophisticated in their thinking about and analysis of cross-cultural issues present in diverse classroom settings" [29, p. 49]. In addition to Cushner and Brislin, other sources of critical incidents are listed in Fowler and Blohm [23].

10.4.4 Open Stories

The openness of stories is relative to the lack of open interpretation in other types of narratives mentioned earlier. As Johnson Sheehan and Flood note about case studies, they are closed because "the rhetorical situation for the case is almost wholly contained within the assignment itself" and students are asked to respond to an identified problem or issue "within a set of guidelines" [24, p. 21]. This same closed-ness, again a relative concept, applies to critical incidents and cross-cultural dialogs. Stories, by contrast, foreground the fuzziness of their narratives, hesitate to define problems or guidelines, and instead encourage readers to explore issues that are not apparent in the narratives and to seek their own solutions.

Open stories are still rare in intercultural technical communication textbooks. Yu and Savage's [30] recent book is probably the first that uses stories to teach the complexity and reward of intercultural technical and professional communication. These stories are based on the experiences of technical communicators who come from a range of cultural and professional backgrounds and who work or have worked in diverse intercultural and multicultural contexts. These stories address important topics such as globally dispersed teams, translation, and localization. Additional readings accompany each story to provide readers with more information regarding the relevant cultural contexts and factors, and discussion questions help students recognize potentially important elements in the story and encourage them to explore the story from diverse angles. An open approach, Yu and Savage argue, helps students defy simple and singular solutions to the challenges in intercultural communication. This approach also enables students to consider the various and complex contextual elements in each interaction, and it allows them to get at a more sophisticated understanding of culture.

These various cultural narratives, as I discuss in the next section, provide an important platform for technical and professional communication teachers to combine the functionalist and interpretive paradigms into a transdisciplinary approach. Upon this platform, they can teach cultural heuristics as well as address extracultural factors. This, I believe, is ultimately a more effective approach to teaching culture and communication.

10.5 Implement the Transdisciplinary Approach: Teach Cultural Heuristics Through Narratives

Whether it is critical incidents or open stories, cultural narratives highlight intercultural communication misunderstandings or conflicts. Students can, in turn, meaningfully discuss these misunderstandings or conflicts through cultural heuristics. Cushner and Brislin [29], Storti [27, 28], and chapters in Bosley [26], for instance, draw upon Hall, Hofstede, and Trompenaars in their analyses. While Yu and Savage [30] pursue open-ended interpretations and omit analyses, they encourage readers to consider heuristics through additional readings and discussion.

At the same time, teachers can use the contextual details in the narrative to complicate or question heuristic explanations. Through contextual details, teachers can guide students to examine additional cultural and extracultural factors such as people's emotions and work roles, socio-economic factors, political factors, and national histories. These examinations can help students avoid generalization and arrive at a more authentic and sophisticated understanding of culture. Narratives are also interesting and engaging to read with their depictions of believable characters and events. This engagement, in my experience, often prompts students to relate classroom learning to life experience and encourages them to share their own stories about intercultural communication.

In the next section, I offer an extended example to illustrate how teachers can use a case study to meaningfully teach heuristics and other cultural and extracultural knowledge. I want to point out that teachers do not always have to use existing cultural narratives but can create their own narratives to meet particular objectives for a course or curriculum. Guidelines on how to write various types of cultural narratives can be found in Storti [27, pp. 133–136], Fowler and Blohm [23, p. 60], Cushner and Brislin [29, pp. 55–56], and the introduction to Yu and Savage [30].

10.5.1 An Extended Example of a Case Study for Teaching Heuristics

The example presented here is based on a case study entitled "Playing the name game" that appears in Markel's *Technical Communication* [25, pp. 39–40]. I chose this case because it is a carefully written cultural narrative and because it is cited in Hunsinger's [11] critique of the heuristic approach. By revisiting this case, I hope to show that alternative readings of the case are possible and that this and similar cultural narratives can bring out the value rather than constraints of cultural heuristics.

"Playing the name game" features Denise McNeil, a 29-year old American computer scientist and founder of the 2-year-old McNeil Informatics. In the case, McNeil competes with several vendors for a project from Crescent Energy, a 40-year-old Saudi oil-refining company that has ties with the Saudi royal family. We learn that Crescent's professional and managerial staff is all male and that women, most of them related to the owner of the company, work as support and clerical staff. We also learn that at McNeil Informatics, women hold a number of top positions.

During the initial briefing between the representatives from Crescent and the bidding companies (all of whom are male except for Denise McNeil), McNeil grew concerned that Crescent might hold gender prejudice against her and, by extension, her company. She also worried that her lead engineer, who has a Jewish last name, might invite ethnic bias from Crescent. McNeil wondered if she should conceal these facts in her proposal in order to stand a chance at winning the project.

Hunsinger [11] expresses concerns about how culture and cultural aspects are portrayed in this case. He argues that while its portrayal of the Saudi cultural practices is not necessarily inaccurate, it creates "an us-versus-them scenario in which the student's role is to deal with odd, foreign, and (from a liberal Western perspective) negative cultural customs" [11, p. 45]. Extracultural factors and intertextual connections such as the oil economy and the second Iraq war, Hunsinger argues, are excluded from the case [11, p. 45].

Hunsinger's above concerns stem from the interpretive paradigm. He objects to the depiction of the Saudis even though that depiction is not necessarily inaccurate. For functionalist scholars, this logic would be puzzling: if something is not inaccurate (and therefore accurate or potentially so given current evidence), why not present it? These conflicting paradigms, as I demonstrate below, can and should be combined.

An important detail in the case is that McNeil is uncertain of Crescent's prejudice. We read that Mr. Fayed, the team leader from Crescent, "smiled slightly" when he shook hands with McNeil and commented that "he did not realize McNeil Informatics was run by a woman" [25, p. 40]. McNeil "did not know what to make of his comment" but nonetheless had "a strong impression that the Crescent representatives felt uncomfortable in her presence. During the break, they drifted off to speak with the men from the other six vendors...." [25, p. 40]. We also learned that McNeil started her company because of the gender and ethnic prejudices she had encountered in the past (presumably in the United States). And toward the end of the case, when McNeil phoned her mentor (referred to here as "Jane Smith") for advice, Smith questioned McNeil's assumption and asked for evidence that Crescent was biased.

Faced with these narrative details, students, given their individual experiences, personalities, and impressions, are likely to make different interpretations and ask different questions—especially if teachers encourage them to do so. Were Mr. Fayed and his fellow representatives indeed harboring gender prejudice with their smile, comment, and action? Or did they lack experience working with strong female figures and were trying to cover up their own anxiety through awkward smiles, offshoot comments, and physical distance? If so, why would they lack this experience? What are the accepted norms for interacting with women in Saudi Arabia's workplace? Is there evidence that Crescent objected to women, not just Saudi women but women everywhere, holding top positions in the workplace? Is it possible that McNeil's suspicion was influenced by the prejudices she had encountered in her own country? What might those prejudices be?

These and other reasonable speculations can help students see culture and intercultural interactions as complex experiences rather than a simplistic us-versus-them, good-versus-bad binary. They also create the occasion to effectively teach cultural heuristics. One heuristic dimension relevant to this case is Hofstede's *uncertainty avoidance*, which, as mentioned earlier, describes the extent to which cultures avoid uncertain situations. After describing what this demonism measures, teachers can explain that according to heuristic studies, Saudi Arabia scores higher than the United States on uncertainty avoidance [31]. Extracultural factors can then be discussed to situate this heuristic dimension in context. For instance, Saudi Arabia's tendency towards high uncertainty avoidance can be traced to the Jabria school of Islam, the Islamic tribal-family traditions, and the legacy of the Ottoman Empire and colonial bureaucracies [32]. At the same time, this tendency is not static: factors such as "[o]il revenues, a rising new middle class, and improved education, health care, communications, and mobility" and "increased interaction with industrial cultures" continue to shape and shift Saudi Arabia's cultural tendencies [32, p. 17].

From this point, students can be guided to discuss how *uncertainty avoidance* might have figured into the case. Given the very different profiles of Crescent and McNeil Informatics, students might speculate that Crescent showed less interest in Denise McNeil because they had never worked with small start-ups and thereby saw McNeil as an unwelcome, unknown risk. Others might speculate that Crescent was not accustomed to working with female leaders so they perceived working with her as an unwelcome, unknown risk. There can be other speculations but no prescribed answers. In essence, a heuristic approach does not automatically bring with it a bias in interpreting culture; rather, it provides a perspective of what may be causing a cultural situation and creates a foundation for further exploration and examination.

At this point, a discussion of Saudi women's role in the workplace becomes relevant. A number of points can be brought into the discussion. Saudi Arabia, in recent studies, scores the lowest among Arab countries in the percentages of women in the workforce, and fewer of these women are likely to hold leadership positions [31, p. 283]. According to Islam, women's primary role is in the family and their professional role cannot interfere with their family role. But once again, situations change, if slowly, as more Saudi women join the workplace and even take on leadership positions in recent years.

Teachers should discuss these factors, as long as they do not encourage the false us-versus-them perception. When it comes to career women, the teaching of Islam and the Saudi perception is not *that* different from the orthodox teaching of Christianity and the traditional American perception. The narrative detail in the case—that McNeil had first experienced gender bias possibly in the United States—is also useful to prevent this binary thinking. We should also remind students that we cannot judge the rights of women (or many other things for that matter) in one country using the purported universal benchmark of another country.

Another heuristic dimension that can be taught with this case is Trompenaars's *Achievement-Ascription*, which, as mentioned earlier, describes the extent to which cultures accord status based on one's achievement or background. Some students might already question that, although McNeil Informatics is a start-up, it has, as the case suggests, "done several projects of this type in the past year" [25, p. 40]. Should not such achievements alleviate Crescent's anxiety? The *Achievement-Ascription* dimension helps address this question.

After explaining what this dimension measures, teachers can introduce that according to heuristic studies, the United States is a culture where status tends to be accorded based on (recent) achievements whereas in Saudi Arabia, status tends to be ascribed to backgrounds, and family background, in particular, is an important source of ascription [6]. For ascribing cultures, recent achievements are not guarantees of future success. Instead, ascribed status is more reassuring because it is time-tested. From this explanation, students can be guided to discuss how this factor may have figured into the case: Maybe Crescent, a company with loyal connections, showed less interest in McNeil because her company, unlike the other vendors, has no background to recommend it? Or, maybe being a 29-year old, McNeil is much younger than the other vendor representatives and this lack of seniority does not sit well with Crescent?

This is also an opportune moment to remind students that heuristic dimensions are statistic approximations. Functionalist scholars do not conceive them as binary extremes and interpretive scholars certainly should not interpret them as such. The *Achievement-Ascription* dimension describes two stages of a continuous circle:

> Those who 'start' with ascribing usually ascribe not just status but future success or achievement and thereby help to bring it about. Those who 'start' with achievement usually start to ascribe importance and priority to the people and projects which have been successful. Hence all societies ascribe and all achieve after a fashion [6, pp. 117–118].

Japanese companies, for instance, might accord status to senior managers, but at the same time, they "spend much on training and in-house education to ensure that older people actually are wiser for the years they have spent in the corporation and for the sheer number of subordinates briefing them" [6, p. 112].

One more heuristic dimension that can be taught through this case is Trompenaars's *Neutral-Affective*, which, as mentioned earlier, describes the extent to which cultures express emotion. After describing what this dimension measures, teachers can explain that according to heuristic studies, people from the United States "tend to exhibit **emotion, yet separate** it from 'objective' and 'rational' decisions" whereas Saudi Arabia, together with some other Arab cultures, are more comfortable expressing feelings at work [6, p. 74]. As Ali [33] writes, Arab negotiators are more likely than their Western counterparts to appeal to emotions and feelings. Based on this understanding, students can be guided to discuss how this dimension may have figured into the case: Is it possible that Crescent showed less interest in McNeil because during their briefing, Denise, eager to prove her company, focused on presenting facts and numbers of her company's past success? Is it possible that other vendors showed more enthusiasm or used emotional appeals?

This discussion about *Neutral-Affective* can be particularly relevant when it comes to writing the project proposal in the case. Should McNeil carry her rhetorical assumption to the writing of the proposal, she will hurt her company's chance at winning the project. Students should also be encouraged to relate this heuristic knowledge (as well as the others described earlier) to their own experience: Do they, as heuristic studies would predict, identify with neutral or affective emotions? Do they have personal narratives that support or contradict this prediction? And what can they learn from such reflections?

> Teachers can use cultural narratives to integrate functionalist and interpretive approaches, to critically teach heuristics, and to examine unique contexts in intercultural interactions.

The previous example illustrates how teachers can use cultural narratives to integrate functionalist and interpretive approaches, to critically teach heuristics, and to examine unique contexts in intercultural interactions. This transdisciplinary approach addresses Hunsginers' [11] critique that the heuristic approach neglects the teaching of extracultural factors. It also effectively realizes his suggestion that technical and professional communication teachers and scholars supplement their intercultural work carried out under the heuristic approach. And for engineering students, this transdisciplinary pedagogy can be engaging, informative, and relevant to both their personal experiences and disciplinary epistemologies.

10.6 Potential Limitations: How to Select Quality Cultural Narratives

Teaching cultural heuristics through narratives is not without limitations. As Corbett writes about the case study option, it "often treats intercultural communication as an intellectual problem which can be solved through a rational dialog, and this limits its ability to enable students to actually cross cultural borders" [34, p. 415]. Instead, Corbett advocates a praxis approach that encourages students "to engage in intercultural experiences and to then analyze their problems with intercultural communication" [34, p. 412]. The praxis approach exposes students to more authentic intercultural contexts with their political and social conflicts, but as Corbett cautions, these conflicts can be risky and must be fully understood and carefully managed before students are engaged in praxis projects.

When it comes to teaching intercultural communication to engineering students, teachers already face many constraints such as the lack of teaching materials, teacher training, and curricular space [1]. In this context, praxis projects may prove too unstructured and time-consuming. Certainly, teaching heuristics through narratives does not conflict with the praxis approach, and teachers should be encouraged to use either or both as it best suits their teaching objectives and priorities.

The constraints that are more central to this pedagogy come from the narratives themselves. To avoid teaching heuristics as binaries and allow the discussion of extracultural factors in situated contexts, we need quality narratives. As Fowler and Blohm note, critical incidents of intercultural communication "need to be carefully written, revised, or selected to make [their] desired point" [23, p. 59]. Similarly, case studies need to be carefully created "to guard against stereotyping or over generalizing" [23, p. 58].

Ineffective cultural narratives, such as the mini case from Chen and Starosta [35], might come off as "the Western master narrative of cultural exchange" [14, p. 15]. Although Chen and Starosta call it a "mini case," this particular narrative is a critical incident from Cushner and Brislin [29] (and the first edition of Cushner and Brislin). Titled "A Secretary's Work is Never Easy," the incident features a Mrs. Jane Simpson, who "enjoyed her job as departmental secretary in a large, well respected university in the United States" and "enjoyed trying to be helpful to students" [29, p. 198]. One day, Mrs. Simpson faced an Indian student who came in and demanded attention to his various problems. The student "never used words such as 'please' and 'thank you,' talked in a tone reminiscent of a superior talking to subordinates, and gave orders to Mrs. Simpson" [29, p. 198]. Mrs. Simpson could not contain her anger and went to see the department chair to ask never to work with this student again.

This incident is, arguably, not the best in Brislin and Cushner [29], because it fails a number of important criteria:

- *First, Balance Character Development* In this incident, Mrs. Simpson is clearly the protagonist and given the most character development: we learn about her work, her attitude toward work, her feelings and emotions, her internal struggle, and her solution to the incident. By contrast, we know next to nothing about the Indian student as a person, which can be quite relevant to this interaction (for instance, we need to know how long he had been in the United States or at this university). All we are told is that he had problems and was demanding attention to those problems. This one-sided character development constructs an alienated cultural Other and prevents reader empathy.

- *Second, Allow Ambiguity* This incident, in and of itself, leaves little room for suspicion or alternative readings. It presents, matter-of-factly, that the Indian student was behaving in an unacceptable manner (according to US standards, that is) whereas Mrs. Simpson was righteous. This rather absolute description obscures the fact that perceptions of intercultural communication, or any communication, are subjective. Meta-discourses such as "Mrs. Simpson *felt* that the student was using a demanding tone" or "Mrs. Simpson *thought to herself*, 'I don't' like that tone'" could have helped to acknowledge subjectivity and create ambiguity. Narrative details, too, can encourage alternative readings. For instance, when Mrs. Simpson went to see her department chairperson, the chairperson might be described as sympathetic yet cautious in accepting Mrs. Simpson's cultural reading. It should be added that the original incident in Cushner and Brislin comes with four interpretations of the incident, which Chen and Starosta [35] did not include in their textbook. These different interpretations do help, somewhat, in encouraging alternative readings of an otherwise unambiguous incident.

- *Third, Balance Drama and Subtlety* One strength of narratives as a teaching method is their ability to engage students. Part of that ability comes from their use of dramatic details. Critical incidents, in order to create critical moments in a very limited space, often have to rely on drama. But too much drama, especially when it is based on negative cultural stereotyping or one-sided character development, can get in the way of productive instruction. Balance between drama and the kind of subtlety used in cross-cultural dialogs is needed.

- **Fourth, Reduce Value Judgment** One's values are integral to his or her cultural identities, so it is next to impossible to suspend value judgment when writing cultural narratives, especially when the purpose of those narratives is to depict cultural misunderstandings or conflicts. But to the extent possible, narratives can be more instructive if they reduce value judgment. As Yu and Savage [30] suggest, this can be done by distancing the protagonist's voice from the Foucauldian author voice. The protagonist in a narrative can present to the world his or her value judgment—which may be incomplete or biased—but this judgment should not appear to be endorsed by the narrative author. Equal character development and ambiguities help to achieve this objective. When we get to hear the cultural Others' feelings and concerns or are encouraged to suspect our own cultural bias, we will be less likely to judge. In addition, narratives should avoid labeling characters but describe what they did, said, thought, or felt to allow reader interpretation. Brislin and Cushner [29] actually did so in this incident, avoiding judgmental words such as "rude" or "patient."
- **Finally, Create Narratives Beyond the Euro-American Experience** This incident, as many other existing narratives about culture and intercultural interactions, focuses on the American (or Euro-American) experience, especially their encounters with the Rest of the World. According to Brislin and Cushner, there are logistic reasons for this:

> most cross-cultural research continues to be done by Euro-Americans or by people who received their training in the United States, Europe, Australia, or New Zealand. A great many sojourners come from these parts of the world, and highly industrialized nations are also the recipients of many sojourners [29, p. 52].

And when these narratives are used in Euro-American educational contexts, it makes sense that the Euro-American experiences are emphasized. But to reach a most inclusive and ethical instruction of culture, teachers and writers of cultural narratives should deliberately include more voices in their work and disrupt the artificial Euro-American versus Rest-of-the-World binary.

10.7 Conclusion

Competence in intercultural communication is invaluable to today's engineering students and professionals. The teaching of this competence, however, is never easy. The topic is inherently complex, is subject to divergent paradigmatic interpretations, and is not part of the traditional engineering curricula. In this chapter, I described a pedagogical approach that embraces both the functionalist and interpretive paradigms: teaching cultural heuristics through narratives. This transdisciplinary approach can give us the best of both worlds, allowing structured learning of cultural dimensions that attempt to explain diverse intercultural interactions as well as the exploration of extracultural factors in situated contexts. It is engaging and can be particularly appealing to engineering students given their disciplinary training. I hope intercultural technical and professional communication teachers will refine this approach through the creation of quality narratives and lessons learned in the classroom.

References

1. H. Yu, "Integrating intercultural communication into an engineering communication service class," *IEEE Transactions on Professional Communication*, vol. 54, no. 1, pp. 83–96, 2011.
2. Edward T. Hall Associates, *Books*, 2012. Available at: http://www.edwardthall.com/books/ (accessed August 2, 2015).
3. E. T. Hall, *Beyond Culture*. Garden City, NY: Anchor Press/Doubleday, 1976.
4. G. Hofstede, *Culture's Consequences: Comparing Values, Behaviors, Institutions, and Organizations across Nations*, 2nd Ed. Thousand Oaks, CA: Sage, 2001.
5. T. Fang, "A critique of Hofstede's fifth national culture dimension," *International Journal of Cross Cultural Management*, vol. 3, no. 3, pp. 347–368, 2003.
6. F. Trompenaars and C. Hampden-Turner, *Riding the Waves of Culture: Understanding Cultural Diversity in Global Business*, 2nd ed. New York: McGraw-Hill, 1998.
7. P. W. Cardon, "A critique of Hall's contexting model: A meta-analysis of literature on intercultural business and technical communication," *Journal of Business and Technical Communication*, vol. 22, no. 4, pp. 399–428, October 2008.
8. E. Schilling and A. Kozin, "Migrants and their experiences of time: Edward T. Hall revisited," *Forum: Qualitative Social Research*, vol. 10, no. 1, Art. 35, January 2009. Available at: http://www.qualitative-research.net/index.php/fqs/article/view/1237 (accessed August 1, 2015).
9. P. Signorini, R. Wiesemes, and R. Murphy, "Developing alternative frameworks for exploring intercultural learning: A critique of Hofstede's cultural difference model," *Teaching in Higher Education*, vol. 14, no. 3, pp. 253–264, June 2009.
10. R. Yeh and J. J. Lawrence, "Individualism and Confucian dynamism: A note of Hofstede's cultural root to economic growth," *Journal of International Business Studies*, vol. 3, pp. 655–669, 1995.
11. R. P. Hunsinger, "Culture and cultural identity in intercultural technical communication," *Technical Communication Quarterly*, vol. 15, no. 1, pp. 31–48, 2006.
12. E. Weiss, "Technical communication across cultures: Five philosophical questions," *Journal of Business and Technical Communication*, vol. 12, no. 2, pp. 253–269, 1998.
13. L. Beamer, "Finding a way to teach cultural dimensions," *Business Communication Quarterly*, vol. 63, no. 3, pp. 111–118, 2000.
14. D. Munshi and D. McKie, "Toward a new cartography of intercultural communication: Mapping bias, business, and diversity," *Business Communication Quarterly*, vol. 64, No. 3, pp. 9–22, 2001.
15. B. McSweeney, "Hofstede's model of national cultural differences and their consequences: A triumph of faith – a failure of analysis," *Human Relations*, vol. 55, no.1, pp. 89–118, January 2002.

16. D. Williamson, "Forward from a critique of Hofstede's model of national culture," *Human Relations*, vol. 55, no. 11, pp. 1373–1395, Nov 2002.
17. G. Burrell and G. Morgan, *Sociological Paradigms and Organisational Analysis: Elements of the Sociology of Corporate Life*. Burlington, VT: Ashgate Publishing Company, 1979.
18. B. Thatcher, "Editor's introduction to first edition: Eight needed developments and eight critical contexts for global inquiry," *Rhetoric, Professional Communication and Globalization*, vol. 1, no. 1, pp. 1–34, 2010.
19. A. D. Figueiredo, "Toward an epistemology of engineering," in *2008 Workshop on Philosophy and Engineering, The Royal Academy of Engineering*, London, November 10–12, 2008, pp. 94–95.
20. A. Penrose and S. Katz, *Writing in the Sciences: Exploring Conventions of Scientific Discourse*. New York: Longman, 2010.
21. D. A. Jameson, "Telling the investment story: A narrative analysis of shareholder reports," *Journal of Business Communication*, vol. 37, no. 1, pp. 7–38, 2000.
22. J. Perkins and N. Blyler, "Introduction: Taking a narrative turn in professional communication," In *Narrative and Professional Communication*, J. Perkins and N. Blyler, Eds. Stamford, CT: Alex Publishing, 1999, pp. 1–34.
23. S. M. Fowler and J. M. Blohm, "An analysis of methods for intercultural training," In *Handbook of Intercultural Training*, D. Landis, J. Bennet, and M. J. Bennett, Eds, 3rd edition. Thousand Oaks, CA: Sage, 2004, pp. 37–84.
24. R. Johnson-Sheehan and A. Flood, "Genre, rhetorical interpretation, and the open case," *IEEE Transactions on Professional Communication*, vol. 42, no. 1, pp. 20–31, March 1999.
25. M. Markel, *Technical Communication*. Boston, MA: Bedford/St. Martin's, 2010.
26. D. S. Bosley, *Global Contexts: Case Studies in International Technical Communication*. Boston, MA: Allyn & Bacon, 2001.
27. C. Storti, *Cross-Cultural Dialogues: 74 Brief Encounters with Cultural Difference*. Yarmouth, ME: Intercultural Press, 1994.
28. C. Storti, *Old World/New World: Bridging Cultural Differences: Britain, France, Germany and the U.S.* Yarmouth, ME: Intercultural Press, 2001.
29. K. Cushner and R. W. Brislin, *Intercultural Interactions: A Practical Guide*, 2nd ed. Thousand Oaks, CA: Sage, 1996.
30. H. Yu and G. Savage, *Negotiating Cultural Encounters: Narrating Intercultural Engineering and Technical Communication*. Hoboken, NJ and Piscataway, NJ: Wiley & IEEE press, 2013.
31. D. M. Noer, C. R. Leupold, and M. Valle, "An analysis of Saudi Arabian and U.S. managerial coaching behaviors," *Journal of Managerial Issues*, vol. 19, no. 2, pp. 271–287, 2007.
32. A. J. Ali, "Management theory in a transnational society: The Arab's experience," *International Studies of Management & Organization*, vol. 20, no. 3, pp. 7–35, 1990.
33. A. J. Ali, "Decision-making style, individualism, and attitudes toward risk of Arab executives," *International Studies of Management & Organization*, vol. 23, no. 3, pp. 53–73, 1993.
34. J. Corbett, "From dialog to praxis: Crossing cultural borders in the business and technical communication classroom," *Technical Communication Quarterly*, vol. 5, no. 4, pp. 411–424, 1996.
35. G. Chen and W. J. Starost, *Foundations of Intercultural Communication*. Boston, MA: Allyn & Bacon, 1998.

11

Assessing Intercultural Outcomes in Engineering Programs

Darla K. Deardorff
Duke University

Duane L. Deardorff
North Carolina State University

This chapter explores the assessment of intercultural learning outcomes within engineering programs and does so by discussing several key frameworks individuals can use to perform such assessment. The authors begin with a general overview of assessment processes within intercultural learning and key principles associated with good assessment practices. Examples of intercultural outcomes assessment within engineering programs are highlighted. A checklist is provided for readers to use in developing assessment plans for engineering programs.[1]

---

[1] Portions of this chapter are adapted from D.K. Deardorff and A. van Gaalen, "Outcomes Assessment in the Internationalization of Higher Education" in *The Sage Handbook of International Higher Education*, D.K. Deardorff, H. de Wit, J. Heyl, and T. Adams, eds. Thousand Oaks, CA: Sage, 2012, pp. 167–188.

Teaching and Training for Global Engineering: Perspectives on Culture and Professional Communication Practices, First Edition. Edited by Kirk St.Amant and Madelyn Flammia.
© 2016 The Institute of Electrical and Electronics Engineers, Inc. Published 2016 by John Wiley & Sons, Inc.

11.1 Introduction

The Georgia Tech website "Welcome to the International Plan" [1] reflects what universities and colleges are becoming increasingly aware of—that companies are seeking graduates who are able to

- Recognize how their discipline is practiced in an international context
- Function effectively in multinational work environments
- Assimilate comfortably into different world cultures
- Assimilate easily into diverse communities and work environments.

The website articulates recent findings gathered from a wide range of US engineering colleges and programs. This collected data reveals that engineering companies are interested in graduates who have more than just content knowledge and technical skills. Rather, prospective employers are interested in hiring graduates who can also demonstrate broader "professional skills"—skills that enable individuals to be successful in diverse settings with persons from a range of backgrounds. Such competencies are becoming valued as much as those technical skills as employers increasingly seek out prospective employees who can successfully work in and navigate through diverse work environments in both their own home countries and abroad. Yet, how do we know that students have the necessary capabilities to succeed in diverse contexts? What is the evidence that students have these global skills needed by engineering companies? These questions lead directly into the topic of this chapter: assessment.

Given the increased call for accountability within higher education, outcomes assessment has emerged as a key focus both within the overall sector and for individual educators. Moreover, such a focus seems to be here to stay [2–9]. At the same time, organizations such as the Lumina Foundation in the United States have generated increased focus on outcomes through initiatives such as the Tuning Project. (The Tuning Project is inspired by the European Union's Tuning Educational Structures in Europe Project, which has become a worldwide effort known simply as "Tuning" and is an attempt to develop specific content outcomes for different disciplines.[2])

As engineering programs become more global in nature, it becomes increasingly necessary for them to provide quality assessment of the learning that occurs in those programs. The Accreditation Board for Engineering and Technology (ABET) criteria [10] previously assessed engineering programs specifically address global competence, as does a 2004 report from the National Academy of Engineering [11]. Within engineering and engineering education in particular, there is also substantial literature citing the importance of global competence within the field. Moreover, engineering scholars note that future engineers will need to be skilled at working with colleagues from different national and cultural backgrounds if they wish to succeed professionally [12, 13].

This chapter provides readers with a foundation for understanding the assessment process as it relates to intercultural outcomes by discussing a tool educators can use

---

[2] See http://www.relint.deusto.es/TuningProject/background.asp for more information on this initiative.

to frame assessment efforts for gauging students' cultural competence. In so doing, the authors also provide a general overview of assessment processes within intercultural learning and examine principles associated with good assessment practices within this particular topic area. The authors also share examples of intercultural outcomes assessment and provide readers with a checklist that can be used in developing related assessment plans for engineering programs.

11.2 An Introduction to the Literature of Outcome Assessments

A holistic approach for thinking about outcomes assessment in a course or program is the program logic model (or "logic model") that will be discussed in further detail in this overall section. This logic model, which starts with the inputs and continues through to the overall impact, can ensure a systematic, aligned, and rigorous approach to assessment efforts both at the student level and at the program level. The model includes the following dimensions:

- Inputs
- Activities
- Outputs
- Outcomes
- Impact

By applying a framework such as the logic model, outcomes assessment can be placed within a more holistic context. Such an approach can ensure a systematic and rigorous approach to assessment efforts.

11.2.1 Thinking About Intercultural Assessment

The framework created by this logic model can be used not only to provide a road map for clarifying intended outcomes, but it can also serve as an analytical tool that leads to lasting change within a course, an overall program, or a larger organization. Aimed at driving such change within an organization, the model is widely used in the private, public, and non-profit sectors that include foundations and even certain agencies within the US federal government.

In creating a "road map," the logic model is a tool that addresses three essential questions that lead to transformational learning and—in this case—*intercultural* learning efforts:

- Where are we going? (Goals)
  The idea is to determine where students should end up in regard to intercultural learning. In many cases, this would be along the lines of "enhanced intercultural competence development." Goals are too broad to be assessed which is why those need to be broken down and addressed in the next question.

- How will we get there? (Outcomes)
  The focus is to determine how students will achieve this goal as well as what the "stepping stones" (objectives) along the way to reach the end goal will be. These objectives are assessable.
- How will we know when we have arrived? (Assessment)
  The notion is to determine what *evidence* will be used to determine the successful achievement of the objectives.

These three main questions can focus efforts on the end goals of intercultural learning within engineering programs. They also help individuals engaged with assessment focus on the *evidence* that these goals have been achieved. That is, within a given course, these questions help identify what exactly students should be able to do interculturally by the end of the course or program. The emphasis is on *change* in the demonstration and the performance of acquired intercultural skills and not on the acquisition of knowledge alone.

11.2.2 Connections to Assessment

To better understand how such factors affect assessment, it is important to review how the specific components of the logic model framework operate by examining them in more detail:

11.2.2.1 Inputs What is needed to achieve the stated goals? Inputs are the resources needed to develop and implement activities that will achieve such goals. Inputs can include staff, faculty, administrators, time, money, partners, facilities, or other resources available within the parameters of the course or program. Inputs can also include employer surveys and focus group results, and alumni surveys. These inputs can then be mapped and assessed in terms of how they might affect the other four components of this model.

11.2.2.2 Activities What are the specific learning activities undertaken to achieve the goals? In the case of intercultural learning within engineering programs, activities can involve learning opportunities that occur through curriculum (e.g., virtual projects and connections, service learning in the community, curricular examples from other cultures—see Section 11.7.1 in this chapter for further discussion on this), education abroad experiences (e.g., Engineers Without Borders), research abroad or in virtual global teams, involvement of faculty abroad, and so on.

11.2.2.3 Outputs Who is involved and is being reached through the activities? Outputs often include participation numbers such as the number of students in an engineering course, an education abroad program, the number of students benefitting from a joint program, the number of joint publications, and so on. A prime example is participation numbers published in the annual *Open Doors* report published by the Institute of International Education [14].

Inputs > Activities > Outputs > Outcomes > Impact

FIGURE 11.1. Logic model components. (This model illustrates how to adapt the components of the program logic model to the internationalization of higher education.)

11.2.2.4 Outcomes What are the *changes* that occur in the learning activity for individuals, programs, departments, and even for the institution? These are the expected or resulting changes of the outputs of an intervention (activities). Outcomes are directly related to the set goals and therefore often identified on a deeper level than outputs—as in the meaning behind the numbers (outputs). Intercultural learning outcomes, which can include short, medium, and even long-term outcomes, are often stated in terms of specific knowledge, skills, and attitudes to be acquired by students. One example of such an outcome statement in an engineering program can be seen in that of the Program in Global Engineering at the University of Michigan, which includes among its main objectives "to appreciate people, culture, and engineering practices of other nations" and "to develop students' capacities for intercultural sensitivity" [15, p. 1].

11.2.2.5 Impact This factor encompasses a number of aspects/questions including: What is the long-term impact (consequences/results) of students' intercultural learning on the educational program and students/alumni—as well as on stakeholders such as staff, business, industry, local community, and international partners? What are the changes in policies, for example? And how has the engineering program itself changed as a result of international partnerships, and of incorporating intercultural learning into the program? Longitudinal studies are often necessary to assess long-term impact. As Deardorff and van Gaalen [16] note,

> One example of such a study is the Dwyer and Peters survey among participants of the Institute for the International Education of Students (IES) study abroad programs. This study is valuable for its use of data over a considerable time period (alumni who participated in these programs between 1950-1999) to assess mid and long term (50 years) impact on the participants' career and personal development. Another impact example is The Study Abroad for Global Engagement (SAGE) project, which ran from 2006-2009. In this project, data were collected in documenting the long-term personal, professional, and global engagement outcomes of study abroad experiences [p. 170].

Figure 11.1 illustrates the basic logic model containing the five basic components, which can be used in this case to map the intercultural elements of an engineering course or program.

Using a framework such as the logic model is often one of the first steps in approaching assessment. Doing so, moreover, helps shift the focus of assessment from reliance on numbers (outputs) to determining the meaning behind the numbers (e.g., the achievement of intercultural learning outcomes, and ultimately, to determining effect of international and global efforts in an engineering program). Of course, no model is perfect, and the limitations of the logic model include the challenge of causal attribution because a variety of factors (e.g., students' backgrounds and prior experiences) might influence outcomes. (For a further discussion on this model, see Deardorff [17, 18].)

11.3 Exploring Some Limitations to Intercultural Assessment Research

In reviewing outcomes assessment from higher education studies conducted in various parts of the world, it is helpful to understand limitations of intercultural outcomes assessment in engineering. This need is particularly acute given that much more research is needed on the topic of intercultural outcomes assessment, especially within the context of engineering. Some research limitations to consider when doing such assessment research include the items covered in this section.

11.3.1 Variation

The variety of engineering programs and courses around the world means there is no universal approach in regard to assessing outcomes given that assessment measures must be aligned with the mission, goals, and objectives of the individual program or the institution in which it is housed.

> Within engineering, there have been attempts to find a common tool to assess intercultural learning [19]. However, not all engineering programs are identical. This factor, coupled with the complexities associated with intercultural learning, means it is not possible for one tool to be able to measure the complexities of intercultural learning.

> It is similarly not possible for one tool to fit the mission, goals, and learning outcomes of engineering programs at multiple institutions of higher education. Leeuw and Vaessen [20], for example, note that no single assessment method is best to address the complexity of the issues at hand.

11.3.2 Self-Orientation

Studies on the effects of education abroad, especially related to intercultural competence development and intercultural learning, are often based on pre/post self-perspective tools as well as on the self-selection of participants. Methodologically, there is a strong preference among administrators for surveys and self-assessment rather than direct or objective measurements or a combination of methods. (This perspective might be founded in the perceived logistical challenges associated with administering a mixed method approach to assessment.) However, direct and indirect methods complement each other and jointly offer a more complete picture than any single method could if used individually. Deardorff [21, pp. 477–491], for instance, found that a multi-perspective method provides a more complete assessment picture of an individual's intercultural competence whereas reliance on one tool alone is insufficient in fully assessing this competence.

11.3.3 Causality

Given the complexities of intercultural learning, most studies can only indicate correlations but not causations. As Leeuw and Vaessen [21] and as van der Vijver and Leung [22] note, directly relating learning interventions (i.e., intercultural service learning, study abroad, specific intercultural workshops, courses, etc.) to desired effects requires knowledge about what the situation would have been without the intervention. Contrary to principles of good research, studies on outcomes and impact within international higher education rarely use control groups, (Exceptions to this practice include the studies done by Sutton and Rubin [23] and those done by Sakurai et al. [24].)

11.3.4 Skills

Studies on the effects of intercultural learning experiences on students generally do not include the backgrounds and baseline skills of these students. In contrast to other more technical engineering skills, intercultural skills are usually not assessed given the perceived ambiguity and nebulousness of these skills. This factor points to the importance of assessing such skills through observation in authentic, real-life settings and not in the more contrived classroom settings.

11.3.5 Focus

Studies on intercultural outcomes show a tendency to focus on conditions for desired outcomes or on the actual outcomes themselves (in the short term) rather than on the process of acquiring intercultural outcomes, which can be as much or more important than actual outcomes [18]. Further, such studies measure short-term impact rather than long-term impact.

11.3.6 Bias

One of the factors influencing the research questions guiding assessment practices is the criteria set forth by the funding agency. This item can introduce a bias that can skew research results. For example, a government-funded agency might require a report that is intended to benefit the citizens of that particular country, and this bias can lead to a national-centric perspective rather than a global perspective. Additionally, the research questions themselves may be culturally biased, with certain assumptions made which can greatly affect results—significantly impacting perceived success (or lack thereof) in an engineering context.

11.3.7 Quality

Outcome assessment studies can vary in terms of quality. It is therefore important to examine the assessment methodologies, data, and analyses associated with a given study in order to determine the strength and quality of it. In many cases, those individuals engaged in intercultural learning assessment might not have a thorough understanding of what makes for rigorous assessment, or even a thorough knowledge of the complexities of intercultural learning (e.g., intercultural assessment needs to go beyond a pre-post application as discussed later in the chapter).

11.3.8 Gaps

Obviously, many of the characteristics of current studies, which include a large majority of case studies, limit the degree to which the related results can be generalized. Furthermore, research studies themselves need to be carefully reviewed and understood in terms of research design, methodology, data, and actual findings. (In intercultural assessment research, for example, it is very important to check alignment of the intercultural assessment tools with the research questions.) Additionally, numerous gaps in the research literature exist in regard to outcomes assessment in engineering education. More studies are therefore needed on the actual assessments of international and intercultural learning outcomes. Such studies need to address these limitations, including cultural biases of both the assessment tools and the methods used.

11.4 Strategies for Quality Assessment of Intercultural Learning Outcomes

Because engineering programs now consider intercultural/global competence to be an important learning outcome for students, we should examine this concept more closely, and an effective point to begin this examination is with a brief note on vocabulary. There are many different terms used to refer to this concept of intercultural competence depending on the discipline and professional field. The most commonly used term in many engineering contexts tends to be *global competence* [25] although *intercultural competence* is also widely used and applies to any individual who interacts with persons from different backgrounds regardless of location [21]. For this reason, this broader term of *intercultural competence* is used throughout the remainder of this chapter.

In terms of assessing intercultural competence, the first step in the overall process involves defining the concept itself and doing so in a way that uses existing literature as a basis for the definition/framework. (This topic will be examined in further detail later in this chapter.) One of the unique aspects of intercultural competence definitions is that they often go beyond knowledge to include skills and attitudes. Thus, engineering courses should focus on more than just knowledge within intercultural learning. So, for example, how do engineering courses foster curiosity (an attitude) in learning how engineers in other cultures approach and address problems? How do courses inspire creativity and open-mindedness? While engineering instructors themselves may not feel as qualified to explicitly address intercultural competence in a course, they can address some of the attitudes and skills related to intercultural competence as discussed in Sections 11.5.3 and 11.5.4 of this chapter.

11.4.1 Establishing Definitions

Most definitions and models of intercultural competence tend to be somewhat general in nature. As a result, once a definition has been determined,[3] it is important to engage

---

[3] A process that often involves a discussion among stakeholders in order to determine the most workable definition within that particular engineering context.

in a process that generates measurable outcomes and indicators within the knowledge, skills, and attitudes areas to be assessed. (In sum, one needs to make sure outcomes and related assessment mechanisms are based on the same definition.) To begin this process, it is best to prioritize specific aspects of intercultural competence and to do so based on the overall mission, goals, and purpose of a course or a program.

The definition that is used for intercultural competence will determine the aspects that will be assessed, and in this case, the intercultural learning that occurs for each student. For example, based on the overall mission of "understanding others' perspectives," related factors might be an essential aspect of intercultural competence to assess. Accordingly, achieving such understanding becomes a stated goal. From that point, one would engage other key persons (e.g., assessment experts on campus, students, other faculty including those in engineering, education, and communication) in a dialogue about the specific measurable outcomes related to this overall goal of "participants' ability to understand others' perspectives" as to the best ways to achieve this goal. These ways of achieving the stated goal thus become the specified outcomes, which will be discussed in more detail later in this section.

11.4.2 Stating Goals and Measurable Outcomes

It is important to spend sufficient time defining intercultural learning goals in such a way that they can be measured regardless of if they were reached or not. Much of the assessment literature, in turn, calls for taking broader goals and developing more specific measurable outcomes statements that can be assessed [3, 4]. In developing specific outcomes statements, individuals can follow the commonly known SMART objectives as follows:

- **Specific**: Is the objective well defined and clear to all stakeholders?
- **Measurable**: In what form will this objective be reached and under what observable conditions? (In other words, what evidence—or performance indicators—will be collected?)
- **Agreed upon**: Do the stakeholders accept this objective?
- **Realistic**: Is assessment possible within the known parameters including availability of resources and time?
- **Time-bound**: Within what timeframe should the objective be reached?

Given that these SMART objectives provide a concrete format one can use to write specific, measurable objectives, this approach means student learning outcomes become more "doable." However, when using this format, it is important to remember what is most important in terms of learning and measure is what matters, not what is easy to measure.

11.4.3 Writing Outcomes

Once specific aspects of intercultural learning have been prioritized (e.g., understanding an engineering problem from others' perspectives), it is time to write measurable outcomes (i.e., objectives) statements that are related to each of the prioritized aspects. In essence, a key part of assessment is to ensure realistic outcomes statements—statements that address the questions of

- Can these outcomes be accomplished within the parameters of the course or program?
- Are these outcomes specifically addressed in the program/curriculum?

In other words, the course or curriculum needs to adequately address and support intercultural learning outcomes through materials, learning activities, and faculty feedback to students. This focus means going beyond a one-time lecture or added intercultural reading.

This situation presupposes faculty members have an in-depth understanding of intercultural development and competence, and that they are able to provide meaningful feedback students can use to guide their continued intercultural development. Outcomes assessment efforts of individual students can later be aggregated to department and even institutional levels in order to determine overall quality of education and achievement of institutional mission. Such an undertaking can, however, be daunting. For this reason, it is helpful to draw on existing resources, including those at the national level, to aid in these endeavors.

Through its work with several higher education institutions across the United States, the American Council on Education [26] provides a list of common intercultural learning outcomes found at the intersection of international and multicultural education. Those outcomes include the ones such as the following statements (examples adapted from the American Council on Education) that need to be tailored more specifically to a particular engineering course, with specific performance indicators developed for each:

- Attitude: Is willing to engage in diverse cultural situations.
- Attitude: Is willing to learn from others who are culturally different from him/her.
- Knowledge: Understands how engineering is viewed and practiced in different cultural contexts.
- Knowledge: Demonstrates knowledge of the connection between local and global issues
- Skills: Uses knowledge, diverse cultural frames of reference, and alternate perspectives to think critically and solve engineering-related problems.
- Skills: Adapts his/her behavior to interact effectively and appropriately with those who are different.

Other resources to use for outcomes statements include several rubrics developed by faculty through the American Association of Colleges and Universities [27], for these rubrics are related to international education and include entries that focus on global learning, civic engagement, and intercultural learning. The latter rubric on intercultural learning is based on the theoretical work of Bennett [28] and Deardorff [21, 29]. As a result, numerous higher education institutions in the United States have adapted and used this rubric to assess actual student work, especially in student portfolios. The key in successfully adapting these rubrics is to align them with a program's existing goals and measurable outcomes, and then to collect evidence that demonstrates the extent to which these outcomes have been achieved.

11.5 Developing an Assessment Plan

Based on research and given the complexity of intercultural competence, an assessment plan involving a multi-method, multi-perspective assessment approach is desired. For example, advocating the use of multiple measures in assessing competence, Pottinger [30] stresses that "how one defines the domain of competence will greatly affect one's choice of measurement procedures" [30, p. 30]. Pottinger also notes that pen and paper assessment tests have been widely criticized due in part to the test format and in part to the limits a paper test places on the complex phenomena being measured [30, pp. 33–34].

11.5.1 Defining Direct and Indirect Measures

Because competence varies by setting, context, and individual, using a variety of assessment methods and tools ensures a stronger measurement as well as more rigorous assessment. In measuring intercultural competence, a variety of assessments include *direct* and *indirect* measures defined here as follows:

- **Direct measures** are actual *demonstrated evidence* of student learning collected during the learning process (such as course assignments, performance, and capstone projects).
- **Indirect measures** are *perceptions* of student learning often collected outside of the learning experience (as collected through surveys, interviews, and focus groups).

Note that over 100 self-perspective tools, which vary by degrees of reliability and validity, have been developed to assess various aspects of intercultural competence in many different and general contexts. (See References 31, 32 for lists of existing intercultural assessment tools.) Most of those tools involve indirect measures and are quantitative in nature. It therefore becomes important to explore *actual evidence of changes* in student learning, beyond students' self-perspectives. This often means combining indirect measures with more qualitative measures.

In using the definition of "*effective and appropriate* behavior and communication in intercultural settings" [29] [18], there is in an inherent implication that measures need to be multi-perspective and go beyond solely the student perspective.

While the student can indicate to what degree he/she has been *effective* in an intercultural setting such as a diverse engineering team, it is only the other person (i.e., the person doing the assessment) who can determine the *appropriateness* of behavior/communication in the interaction itself. For example, a supervisor or a professor might use a specifically developed intercultural rubric to conduct an observation of student performance in a particular situation (e.g., a team project), or a peer (e.g., fellow team member) may be in the best position to indicate the appropriateness of the response in the interaction.

11.5.2 Making Sense of Intercultural Assessment

What does this all mean in assessing intercultural competence? The central concept is that intercultural assessment must involve more than one measurement tool. This chapter has argued that given the complexity of this concept, it would be challenging—if not impossible—for one tool to measure an individual's intercultural competence. Moreover, there are more in-depth questions to address including: "Intercultural competence from whose perspective and according to *whom?*" or "Intercultural competence to what degree?" or "Intercultural competence to what end?" These questions are crucial to explore further in developing appropriate measures of intercultural competence.

Additionally, specific priorities will vary by course, program, and department in regard to aspects of intercultural learning. For example, not all engineering programs will have exactly the same goals and objectives, even when following ABET criteria. Thus, an intercultural assessment tool being used in one course or program might not be appropriate for another course or program if the goals for those are different. This point is crucial given the pressure to find one comprehensive way to measure intercultural outcomes within engineering programs. This pressure leads to homogenization of programs given the unique strengths and specialties of each engineering program and institution. Furthermore, such homogenization does not support the unique learning outcomes of each program.

11.5.3 Defining Intercultural Competence

One of the first steps in assessment involves knowing exactly what exactly is being assessed—in this case, in defining the actual concept of intercultural competence. Too often, the term *intercultural competence* (or other similar terms) is used without a concrete definition. As discussed by Fantini [33], it is essential to arrive at a definition of intercultural competence before proceeding with any further assessment endeavors. In defining intercultural competence, it is important to recognize that over five decades of scholarly effort has been invested in this concept within the United States. (See King and Baxter Magolda [34] for a more complete discussion on this topic.) Such work should thus be considered when developing a working definition of intercultural competence. However, two studies ([29] [32]) show that, in the case of post-secondary institutions, such definitions and scholarly work are often not utilized. Instead, definitions often rely primarily on faculty discussions without any consultation of the literature.

There are countless definitions and frameworks published on intercultural competence. The first study to document consensus among leading intercultural experts, primarily from the United States, on aspects of intercultural competence [29], was

DEVELOPING AN ASSESSMENT PLAN 251

determined through a research methodology called the Delphi technique, an iterative process used to achieve consensus among a panel of experts. The aspects upon which these experts reached consensus were categorized under attitudes (as the fundamental starting point), knowledge, skills, desired internal outcomes, and desired external outcomes. Specifically, this model was derived from the need to assess this nebulous concept and hence its focus on internal and external outcomes of intercultural competence, based on the development of specific attitudes, knowledge, and skills inherent in intercultural competence.

Given that the items within each of these dimensions are still broad, each aspect can be developed into more specific measurable outcomes and corresponding performance indicators, depending on the context. The overall external outcome of intercultural competence is defined as the *effective* and *appropriate* behavior and communication in intercultural situations, which again can be further detailed in terms of indicators of appropriate behavior in specific contexts, in this case engineering contexts in different cultural settings around the world.

11.5.4 Key Considerations

There are several key points to consider in this grounded-theory-based model, and these points have implications for assessment of intercultural competence within engineering contexts. According to this grounded-research study, intercultural competence development is an ongoing process. It thus becomes important for individuals to be provided with opportunities to reflect upon and assess the development of their own intercultural competence over time; this points to the importance of process (and not just results), such as the process of working on a team project—not just in the accomplishment of the project itself. Given the ongoing nature of intercultural competence development, it is also valuable to incorporate integrated assessments throughout a targeted intervention and even beyond, when possible (e.g., longitudinal assessment).

Another point to note is that critical thinking skills play a crucial role in an individual's ability to acquire and evaluate knowledge (in this case intercultural knowledge within an engineering context). This factor means critical thinking assessment, which is already addressed in engineering programs, could also be an appropriate part of intercultural competence assessment.

Within this context, attitudes—particularly respect (which is manifested differently in cultures), openness, and curiosity—serve as the basis of this model and impact all other aspects of intercultural competence development. Addressing attitudinal assessment in engineering programs, then, becomes an important consideration. This factor means engineers and engineering students need to be open and curious in learning about the cultural contexts in which they will be working, as well as understanding more about the cultural backgrounds of their colleagues and how such backgrounds impact behavior within engineering teams.

There was only one aspect agreed upon by all the intercultural experts in this Delphi study [29] as absolutely essential to intercultural competence, and that was the ability to see from others' perspectives. Thus, assessing global perspectives and the ability to understand other worldviews becomes an important consideration as well, particularly in regard to different perspectives on engineering problems and solutions. Another aspect in the model, deep cultural knowledge, entails a more holistic, contextual understanding of that culture, including the historical, political, and social contexts that underscore the importance of engineers going beyond technical knowledge to understand deeper cultural contexts when implementing engineering solutions. For example, understanding the historical and/or cultural significance of the geographic placement of a bridge could impact the success of the project. Thus, any assessment of culture-specific knowledge needs to go beyond the conventional surface-level knowledge of foods, greetings, customs, and so on, although that can certainly be a starting point.

11.5.5 Frameworks and Models of Assessment

Knowledge alone is not sufficient for intercultural competence development. As Bok [35] indicates, developing skills for thinking interculturally becomes more important than actual knowledge acquired. In terms of assessing the more visible aspects of intercultural competence (namely the external outcome of effective and appropriate behavior and communication such as in teamwork or project work), it is important to recognize that appropriateness can primarily be assessed through others' perspectives.

There are other models that have been used to frame aspects of intercultural competence. These include Bennett's Developmental Model of Intercultural Sensitivity [28], King and Baxter Magolda's intercultural maturity model [34], and Hunter et al.'s Global Competence Model [32], which are referenced in some engineering programs. There are many other models and frameworks which purport to define intercultural competence (see Reference 31 for a more thorough discussion). Many of those models, however, are not based on actual research. Regardless, in assessing intercultural competence, it becomes very important to define this concept within the context it will be used. To that end, frameworks such as the one highlighted here can become a key tool in laying the groundwork for assessing intercultural competence. These intercultural competence models can, in turn, help engineering educators specifically identify characteristics of intercultural competence that will help engineers be more successful in global contexts. Such factors can then be prioritized, incorporated into the curriculum, and translated into clear learning outcomes that are actually measured or evaluated through assessment plans.

11.6 Quality Assessment

What makes for quality assessment in intercultural outcomes assessment in engineering programs? Discussed here are key strategies that can lead to such successful assessment efforts.

11.6.1 Defining Intercultural/Global Competence

The idea is to draw upon the related literature to develop a working definition that can be used within the program and prioritizing what is being measured. As previously discussed, clearly defining what is to be assessed, based on the literature and existing theoretical foundations, is crucial as a starting point. Given the lack of clarity around definitions of competence, it is little wonder that some individuals within higher education remain skeptical about how such competence, beyond knowledge, can be assessed. In fact, Bok emphasizes this point by stating, "...many Arts and Sciences professors are reluctant to teach skills" [35, p. 250] whereas engineering professors do focus more on the teaching of actual skills. However, discussions on definitions of competence have relevance in understanding how best to assess this outcome in students, such as through knowledge tests, performance-based assessments or trait instruments. As Palomba and Banta conclude, "the act of demonstrating competence provides an opportunity to measure or assess it, and today there are many compelling reasons to assess the competence of college students and graduates" [4, p. 6]. Defining terms and developing working definitions within a specific program context through discussions and dialogues with key stakeholders can be illuminating in understanding prioritizes as to why such intercultural measures should even be assessed.

11.6.2 Developing a Comprehensive Assessment Plan

Such a plan would need to include the learning goals (destination) and learning outcomes (how to reach the destination) as well as aligned assessment tools/methodologies and include clear logistics for how to implement assessment efforts (including a timeline, personnel, and use of data collected). This approach means an assessment plan would be uniquely tailored to a specific program and not borrowed from another school's program. Also, assessment tools/methods would be aligned specifically to the program's goals and outcomes, instead of copying tools/methods used by other schools. Furthermore, the assessment plan would

- Include a multi-measure approach that is integrated into the course or program (beyond a pre-post approach)
- Ensure feasibility through prioritization (as opposed to trying to assess too much at once)
- Collect baseline data and provide for a control group, if needed
- Ensure use of assessment data collected (before it is collected).

A central factor to keep in mind for such processes is that an assessment plan specifies in advance exactly how data will be used. This factor means that assessment will hopefully include opportunities for providing feedback to students to assist with the continued development of their intercultural competence.

11.6.3 Involving a Team in Assessment Efforts

Assessment involves teamwork, collaboration, and the support of many. To garner that needed support and expertise, it is crucial that more than one person be involved in the assessment process. This approach might mean putting together a team of interested engineering faculty, assessment experts on campus, and even students.

11.6.4 Ensuring Intercultural Learning is Intentionally Addressed and Supported

Intercultural learning does not automatically happen. Furthermore, intercultural learning must go beyond one course or experience. Given that ABET criteria address global competence, it becomes incumbent upon engineering programs to ensure that intercultural competence is intentionally integrated throughout engineering course work.

11.7 Developing Intercultural Competence in Students

What does it mean to infuse intercultural competence and global learning into engineering courses?

> First, it is important to understand that it is NOT merely the inclusion of an international reading in a course or addressing this topic in one lecture. Similarly, it involves more than taking just one" international" course or completing a required general education course on this or a related topic.

Such cursory treatment is far too limited in guiding students through the developmental process of intercultural competence acquisition.

11.7.1 Considering the Curriculum

Intercultural competence needs to be addressed throughout many undergraduate courses, including STEM courses (science, technology, engineering, and mathematics). Moreover, faculty themselves need to understand more fully this concept as they integrate it into the curriculum [36, 37]. This infusion of intercultural competence and global learning into courses includes

- Finding multiple ways throughout a course to bring in diverse perspectives on issues
- Helping students begin to see challenges from multiple cultural perspectives
- Including materials from a broad range of national and cultural perspectives, so as not to rely on scholarship originating from one national context (such as all materials being from US-based engineering scholars)
- Using students' diverse backgrounds within a course
- Requiring students to apply their intercultural learning through a local cultural immersion or education abroad experience (possibly through research, service learning and/or internship)

In considering such factors, it is important to note that merely having an international experience does not necessarily mean that intercultural learning goals and outcomes will be achieved. Rather, there must be careful attention given to how students will be

- Prepared interculturally before the international sojourn
- Guided in their intercultural learning while in country
- Debriefed and how international experiences will be further processed upon return to the home country

If students are not able to go abroad, it becomes even more crucial for faculty to develop meaningful learning activities (perhaps via technology), both in and out of the classroom, which allows students to apply intercultural learning within diverse domestic settings.

11.7.2 Considering Aspects of Non-formal Learning

Beyond formal ways of supporting students' intercultural learning, it is important for engineering programs to explore the non-formal learning routes, especially given the diversity of engineering students in the various programs. There is, for example, great need for programs to develop specific and intentional mechanisms that would bring domestic and international students together in meaningful interactions [38]. Such programs would include

- Providing adequate preparation (e.g., during orientation or through cross-cultural training) for domestic/international student interaction
- Having specific intercultural learning goals for all participants
- Encouraging meaningful domestic/international interactions through relationship-building opportunities, both on campus and virtually.

These opportunities can include programs such as community service, mentor programs, language partner programs, and student clubs.

11.7.3 Using Collected Assessment Data

It is crucial to use the collected assessment data in a number of ways and for a variety of reasons including

- To provide direct feedback to students on their continued intercultural learning
- For program improvement (by aggregating student learning outcomes data)
- To communicate results to stakeholders
- For accreditation purposes (which often focuses heavily on student learning outcomes)

In the assessment field, this approach is generally known as "closing the loop," and it is crucial that any collected data actually be used. If there are instances where data, either quantitative or qualitative, are being collected and not used, then there is no need to expend time, energy and resources in the collection of such data. The most important use of assessment data is for students' continued learning.

11.7.4 Evaluating the Assessment Plan and Process

One area that is easy to overlook is in evaluating how well the assessment plan worked and what could be improved in the future. It is important to evaluate how well the tools/methods worked in collecting the evidence needed and whether the same set of priorities should be continued in the future or reset to include other priorities.

11.8 An Example of Intercultural Assessment

While much research is still needed on quality assessment efforts in intercultural learning, we can draw upon cases of successful assessment practices to consider approaches to addressing these issues. This section presents an overview of two engineering programs that have attempted to address and assess intercultural outcomes. The example programs overviewed here are provided as snapshots of what is currently being done. Thus, these examples do not necessarily exemplify "best practices." They do, however, represent programs that are working toward including intercultural assessment within their curriculum. With these ideas in mind, it is necessary to examine briefly each of these example programs further in terms of how these efforts align with principles of good assessment practice.

11.8.1 Georgia Institute of Technology (Georgia Tech)

A stated goal of Georgia Tech's College of Engineering's Strategic Plan notes the importance of ensuring its graduates are global citizens. To that end, through Georgia Tech's Quality Enhancement Plan, a required part of re-accreditation, the institution developed an "International Plan" which is described as offering

> [A] unique degree-long program that integrates
> international studies and experiences into any major at Georgia Tech. [39, p. 1]

The institution goes on to say that

> The International Plan is a challenging four-year program that works in tandem with an undergraduate's academic curriculum to produce globally competent citizens... Successful completion of the program results in a special "International Plan" designation on the Georgia Tech degree and transcript [39].

Assessment measures used in the International Plan include a language test, a self-perspective inventory (Intercultural Development Inventory[4]) and a capstone course [39].

11.8.2 Purdue University

Purdue's Engineering Strategic Plan provides clear goals for what is desired in engineering students, with three broad goals stated as

Goal 1:
Purdue Engineers will be prepared for leadership roles in responding to the global, technological, economic, and societal challenges of the 21st century.
Goal 2:
We will focus our talent and facilities on research with great potential for expanding the boundaries of science and technology and addressing the global challenges and opportunities of the 21st century.
Goal 3:
Together, WE –Faculty, Staff, and Students – will make the environment in which we work, create, and study, the best in the world for the creative intellect we already have and the talent that will join us [40].

Specifically, within one engineering program at Purdue, it was determined that there are three key areas for assessment that align with objectives under Goal 1, and they are

- Foreign language competence
- General cultural orientation and appreciation
- Global engineering competency
- Related methods used to assess these factors include
- A foreign language test
- An instrument called the Miville-Guzman Universality-Diversity Scale - Short Form (MGUDS-S),
- A 15-item instrument that measures universal-diverse orientate on (UDO), or "an attitude of awareness and acceptance of both similarities and differences that exist among people"
- Written global scenarios that allow engineering students to apply global engineering knowledge

Both the Georgia Tech and the Purdue University examples illustrate the use of multiple assessment measures to assess intercultural learning, as well as alignment of measures

---

[4] IDI was developed by Mitch Hammer based on the Developmental Model of Intercultural Sensitivity of Milton Bennett.

with goals. Other institutions addressing aspects of intercultural learning in their engineering programs include Virginia Polytechnic University, Colorado School of Mines, University of Tulsa, and University of Pittsburgh.

11.9 Assessing Intercultural Outcomes in Engineering Programs

What does this all mean? It is important to have an understanding of quality assessment principles as well as definitions of intercultural/global learning. In the following list, we present questions that can be used as a checklist when embarking on the task of assessing engineering students' intercultural learning:

- Are there clearly stated goals and measurable outcomes, based on mission?
- Have terms such as global competence or intercultural competence been consistently defined based on the literature and tailored to the specific engineering context?
- Are intercultural outcomes supported adequately by learning activities/methodologies/materials?
- Do faculty have an understanding of the complexities of intercultural learning and its development process?
- Are assessment tools/methods aligned with measurable outcomes?
- Do such tools/methods collect actual evidence of changes in student learning and involve a variety of tools/methods?
- Is there an assessment team in place (comprised of assessment experts, stakeholders, faculty including those in other disciplines such as education and communication, and administrators including those from the international office)?
- Is there an assessment plan in place, which includes details as to how assessment data will be used?
- Have baseline data been collected, including from a control group?
- Has the assessment process itself been reviewed?

The ideas presented in this checklist provide educators with a foundation they can use to design and implement a quality assessment process that will help provide the evidence that engineering programs are, indeed, graduating globally ready students.

11.10 Conclusion

In the increasingly diverse and interconnected world of the twenty-first century, engineers not only need technical skills, but also crucial intercultural skills as they work with those from different backgrounds toward larger goals. Thus, it becomes imperative for engineering programs to incorporate and assess intercultural learning outcomes into their courses. The complexities of intercultural learning translate into assessment strategies that require time and effort. In sum, there is no easy solution if intercultural assessment is to be done in a rigorous, quality manner in order to determine students' intercultural learning development. Such quality assessment efforts can be well worth the investment in producing engineers who are capable not only of exemplary technical execution, but who are able to successfully work on diverse teams and in our global society.

References

1. Georgia Institute of Technology, *Welcome to the International Plan.* Available at: http://www.internationalplan.gatech.edu/.
2. R. A. Vorhees, *Measuring What Matters: Competency-Based Learning Models in Higher Education*, San Francisco, CA: Jossey Bass, 2001.
3. A. Driscoll and S. Wood, *Developing Outcomes-Based Assessment for Learner-Centered Education: A Faculty Introduction*, Sterling, VA: Stylus, 2007.
4. C. A. Palomba and T. W. Banta, *Assessing Student Competence in Accredited Disciplines*, Sterling, VA: Stylus, 2001.
5. L. Suskie, *Assessing Student Learning: A Common Sense Guide,* 2nd ed., San Francisco, CA: Jossey-Bass Publishers, 2009.
6. P. Maki, *Assessing for Learning: Building a Sustainable Commitment Across the Institution*, 2nd ed. Sterling, VA: Stylus Publishing, 2010.
7. B. E. Walvoord, *Assessment Clear and Simple: A Practical Guide for Institutions, Departments, and General Education*, San Francisco, CA: Jossey Bass, 2004.
8. M. N. Hundleby and J. Allen, Eds., *Assessment in Technical and Professional Communication*, Amityville, NY: Baywood, 2010.
9. J. Penn, Ed., *Assessing Complex General Education Student Learning Outcomes: New Directions for Institutional Research*, Hoboken, NJ: John Wiley & Sons, 2011.
10. Engineering Accreditation Commission, *Criteria for Evaluating Engineering Programs*, Baltimore, MD: ABET Inc, 2006.
11. National Academy of Engineering, *The Engineer of 2020: Visions of Engineering in the New Century*, Washington, DC: National Academies Press, 2004.
12. G. L. Downey, J. C. Lucena, B. M. Moskal, R. Parkhurst, T. Bigley, C. Hays, B. K. Jesiek, L. Kelly, J. Miller, S. Ruff, J. L. Lehr, and A. Nichols-Belo, "The globally competent engineer: Working effectively with people who define problems differently," *Journal of Engineering Education*, vol. 95, issue 2, pp. 1–17, 2006.
13. J. C. Swearengen, S. Barnes, S. Coe, C. Reinhardt, and K. Subramanian, "Globalization and the undergraduate manufacturing engineering curriculum," *Journal of Engineering Education*, vol. 91, no. 2, pp. 255–261, 2002.
14. Open Doors website, *Institute of International Education (IIE), Inc.* Available at: http://www.iie.org/Research-and-Publications/Open-Doors.
15. E. Dey, S. Pang, M. Mayhew, and M. Eljamal, "Outcomes assessment in international engineering education: Creating a system to measure intercultural development," presented at American Society for Engineering Education Annual Conference & Exposition, Portland, OR, 2005.

16. D. K. Deardorff and A. van Gaalen, "Outcomes assessment in international higher education," in *The Sage Handbook of International Higher Education*, D. K. Deardorff, H. de Will, J. D. Heyl, and T. Adams, Eds. Thousand Oaks, CA: Sage, 2012, pp. 167–190.
17. D. K. Deardorff, "A matter of logic?," *International Educator*, vol. 14, no. 3, pp. 26–31, May/June, 2005.
18. D. K. Deardorff, *Demystifying Outcomes Assessment for International Educators: A Practical Approach*, Sterling, VA: Stylus, 2015.
19. B. K. Jesiek, M. Borrego, and K. Beddoes, "Advancing global capacity for engineering education research (AGCEER): relating research to practice, policy, and industry." *Journal of Engineering Education*, vol. 99, no. 2, pp. 107–119, 2010.
20. F. Leeuw and J. Vaessen, *Impact Evaluations and Development. NONIE Guidance on Impact Evaluation*, Washington, DC: Worldbank, Network of Networks on Impact Evaluation (NONIE). Available at: http://www.worldbank.org/ieg/nonie/guidance.html.
21. D. K. Deardorff, *The Sage Handbook of Intercultural Competence*, Thousand Oaks, CA: Sage, 2009.
22. F. J. Van de Vijver and K. Leung, "Methodological issues in researching intercultural competence," in *The Sage Handbook of Intercultural Competence*, D. K. Deardorff, Ed. Thousand Oaks, CA: Sage, 2009, pp. 404–418.
23. R. Sutton and D. Rubin, "The GLOSSARI project: Initial findings from a system-wide research initiative on study abroad learning outcomes," *Frontiers: The Interdisciplinary Journal of Study Abroad*, vol. 10, pp. 65–82, 2004.
24. T. Sakurai, F. McCall-Wolf, and E. S. Kashima, "Building intercultural links: The impact of a multicultural intervention programme on social ties of international students in Australia," *International Journal of Intercultural Relations*, vol. 34, no. 2, pp. 176–185, 2010.
25. J. M. Grandin and N. Hedderich, "Global competence for engineers," in *The SAGE Handbook of Intercultural Competence*, D. K. Deardorff, Ed. Thousand Oaks, CA: Sage, 2009, pp. 362–373.
26. American Council on Education, *Web Guide: Preparing for assessment*. Available at: http://www.acenet.edu/Content/NavigationMenu/ProgramsServices/International/Campus/GoodPractice/fipse/preparing/index.htm.
27. AAC&U, *VALUE: Valid Assessment of Learning in Undergraduate Education*. Available at: http://www.aacu.org/value/rubric_teams.cfm.
28. M. J. Bennett, "Towards ethnorelativism: A developmental model of intercultural sensitivity (revised)," in *Education for the Intercultural Experience*, R. M. Paige, Ed. Yarmouth, ME: Intercultural Press, 1993, pp. 21–71.
29. D. K. Deardorff, "Identification and assessment of intercultural competence as a student outcome of internationalization," *Journal of Studies in International Education*, vol. 10, no. 3, pp. 241–266, 2006.
30. P. S. Pottinger and J. Goldsmith, *Defining and Measuring Competence*, San Francisco, CA: Jossey-Bass, 1979.
31. B. Spitzberg and G. Changnon, "Conceptualizing intercultural competence," in *The Sage Handbook of Intercultural Competence*, D. K. Deardorff, Ed. Thousand Oaks, CA: Sage, 2009, pp. 2–52.
32. B. Hunter, G. P. White, and G. C. Godbey, "What does it mean to be globally competent?" *Journal of Studies in International Education*, vol. 10, no. 3, pp. 267–284, 2006.

33. A. Fantini, "Assessing intercultural competence: Issues and tools," in *The SAGE Handbook of Intercultural Competence*, D.K. Deardorff, Ed. Thousand Oaks, CA: Sage, 2009, pp. 456–476.
34. P. M. King and M. Baxter Magolda, "A developmental model of intercultural maturity," *Journal of College Student Development*, vol. 46, no. 6, pp. 571–592, 2005.
35. D. Bok, *Our Underachieving Colleges*, Princeton, NJ: Princeton University Press, 2006.
36. E. Brewer and K. Cunningham, *Integrating Study Abroad Into the Curriculum: Theory and Practice Across the Disciplines*, Sterling, VA: Stylus, 2010.
37. B. Leask, *Internationalisation of the Curriculum (IoC) in Action: A Guide*. University of South Australia, 2012. Available at: http://uq.edu.au/tediteach/OLT/docs/IoC-brochure.pdf.
38. C. Glass, R. Wongtrirat, and S. Buus, *International Student Engagement: Strategies for Creating Inclusive, Connected, and Purposeful Campus Environments*, Sterling, VA: Stylus, 2014.
39. Georgia Institute of Technology. *Georgia Institute of Technology's Quality Enhancement Plan: Impact Report*. Available at: http://www.internationalplan.gatech.edu/ https://www.accreditation.gatech.edu/wp-content/uploads/2011/03/QEP-Impact-Report_SACSCOC_March-25-2011.pdf.
40. Engineering as a profession. *College of Engineering, Purdue University*. Available at: http://www.purdue.edu/catalogs/engineering/profession.html.

Biographies

Editor Biographies

Kirk St.Amant is the Eunice C. Williamson Endowed Chair in Technical Communication at Louisiana Tech University and an Adjunct Professor of International Health and Medical Communication with the School of Culture & Communication, University of Limerick. His research interests include medical communication in global contexts, internationalizing online education, and international aspects of online communication.

Madelyn Flammia is an Associate Professor of English at the University of Central Florida in Orlando, Florida US. She teaches both graduate and undergraduate courses in technical communication. Her research interests include international technical communication, global citizenship, and virtual teams. Dr. Flammia is the coauthor of *Intercultural Communication: A New Approach to International Relations and Global Challenges*. She has published articles in *IEEE Transactions on Professional Communication*, *Technical Communication*, and the *Journal of Technical Writing and Communication*, among others. Dr. Flammia has given presentations on intercultural communication and on global virtual teams at professional conferences and for corporate audiences.

Author Biographies

Bethany Aguad
Bethany Aguad is an Associate Technical Communicator at Wyndham Vacation Ownership. A member of the Orlando Central Florida STC Chapter, Bethany has been a Sigma Tau Chi honoree, winner of the Distinguished Student Service Award, and comanaged the chapter's mentoring program pairing professionals with student members. A graduate of the University of Central Florida (UCF), she has served as president of the Future Technical Communicators (FTC) organization, president of the Sigma Tau Delta Zeta Xi

chapter (International English Honor Society), and received the prestigious STC/UCF Melissa Pellegrin Memorial Scholarship for excellence in technical communication.

Brian Ballentine
Prior to completing his Ph.D. at Case Western Reserve University, Brian Ballentine was a Senior Software Engineer for Marconi Medical and then Philips Medical Systems designing user interfaces for web-based radiology applications and specializing in human–computer interaction. He is currently an Associate Professor and the Associate Chair for the Department of English at West Virginia University in the United States. His recent publications have appeared in journals such as *Technical Communication*, *IEEE Transactions on Professional Communication*, *Computers and Composition Online*, and *Across the Disciplines* as well as in several collected editions.

Carol Barnum
Carol Barnum is Professor Emeritus at Southern Polytechnic State University, US, where she directed a graduate program in information design and communication and taught courses in global communication, usability testing, and information design. She is currently cofounder and director of Research for UX Firm. Barnum is also a Fellow of the Society for Technical Communication and a recipient of the Gould Award for Excellence in Teaching Technical Communication and the Rainey Award for Excellence in Research. She is also a recipient of the IEEE Professional Communication Society's Blicq Award for Distinction in Technical Communication. Barnum is the author of six books including *Usability Testing Essentials: Ready, Set...Test!*

Audrey Bennett
Audrey Bennett is an Associate Professor of Graphics in the Department of Communication and Media at Rensselaer and a Mellon Distinguished Scholar at the University of Pretoria, South Africa. She studies the design and implementation of images that effect behavioral and cognitive change. Her publications include "Engendering Interaction with Images" (monograph) and "The Rise of Research in Graphic Design," which introduces the 2006 collection she edited titled *Design Studies: Theory and Research in Graphic Design*. She is coeditor of the "Icograda Design Education Manifesto 2011" and founder of GLIDE, a biennial virtual conference on global interaction in design education.

Pam Estes Brewer
Dr. Pam Estes Brewer directs the Master of Science in Technical Communication Management in Mercer University's School of Engineering, located in Macon, Georgia, US. She has worked as a technical communicator for several software and database engineering companies and as a contractor. She is an Associate Fellow of the Society for Technical Communication (STC), manager of STC's Academic SIG, and a recipient of the Gould Award for excellence in teaching. Brewer also received the 2014 STC President's Award. She is a member of the editorial staffs of *IEEE TPC* and *Technical Communication*. Dr. Brewer's book, *International Virtual Teams: Engineering Global Success*, was published in the IEEE PCS Professional Engineering Communication Series in 2015.

Yvonne Cleary

Yvonne Cleary is a Lecturer in Technical Communication, and Program Director for the MA in Technical Communication and E-Learning at the University of Limerick, Ireland. Dr. Cleary's teaching and research interests include technical communication in Ireland, technical communication pedagogy, virtual teams, and international communication. She has presented her work at conferences in Europe and the United States and has published in leading journals in the technical communication field. She is a 2014 recipient of the CCCC Technical and Scientific Communication Award for Best Article on Pedagogy or Curriculum in Technical or Scientific Communication.

César Correa Arias

César Correa Arias holds a Ph.D. in Science of Education, University of Toulouse, France. He is a full time teacher and researcher at the University of Guadalajara, since 2003, and an Associated Professor at the École des Hautes Études en Sciences Sociales, in Paris, France, since 2012. The topics he focuses on in his teaching and research include public policies in higher education and curriculum and didactics. He is member of the National Research Council of Mexico; the Sciences Academy of Jalisco, Mexico; the Latin American Sociology Association; and the Tolousaine Researchers Association, France. He has also written different articles in international journals and published books about contemporary higher education.

Darla K. Deardoff

Darla K. Deardorff is Executive Director of the Association of International Education Administrators (AIEA), based at Duke University, where she is also a Research Scholar. In addition, she holds faculty positions at numerous institutions including Nelson Mandela Metropolitan University (South Africa), Meiji University (Japan), and Harvard's Future of Learning Institute. Editor of *The Sage Handbook of Intercultural Competence* (Sage, 2009), she has published widely on international education and cross-cultural issues, including her most recent book *Demystifying Outcomes Assessment for International Educators* (Stylus, 2015). Founder of ICC Global, she is regularly invited to give talks, trainings, and workshops around the world.

Duane L. Deardoff

Duane L. Deardorff is a Physics Professor at the University of North Carolina at Chapel Hill, US, where he specializes in assessing students' understanding of physics and in developing curricula and class environments to enhance student learning. He has presented numerous talks on students' understanding of physics, assessment strategies, physics lab innovations, and the development and implementation of integrated lecture/laboratory courses. He is a recognized expert in measurement uncertainty, and his doctoral research on this topic was conducted through a National Science Foundation grant to Japan and Korea. He also has experience teaching physics to engineering students at North Carolina State University.

Helen M. Grady
Helen M. Grady is a Professor in the School of Engineering, Mercer University, and Chair of the Department of Technical Communication. She has taught over 20 different undergraduate and graduate technical communication and entry-level engineering courses since 1991. Prior to joining Mercer, she managed a technical publications division for a Fortune 100 corporation in Research Triangle Park, NC, US for 10 years. Grady is a senior member of IEEE, serves on the IEEE Professional Communication Society (PCS) Administrative Committee, and received the Schlesinger Award for service to PCS in 2006. She is also an STC Fellow and a recipient of the Jay R. Gould Award for teaching excellence in technical communication.

Steven Hammer
Steven Hammer is an Assistant Professor of Communication and Digital Media at Saint Joseph's University in Philadelphia, PA, USA. His teaching and research investigates the role of digital technologies and platforms on the creation, distribution, and consumption of information and creative work.

Bruce Maylath
Bruce Maylath, Professor of English at North Dakota State University, teaches courses in linguistics and international technical communication. His current research takes up translation issues in professional communication and has appeared in *connexions*, *IEEE-Transactions in Professional Communication*, *Journal of Business and Technical Communication*, *Research in the Teaching of English*, and *Technical Communication Quarterly*, among others. His coedited anthologies include *Approaches to Teaching Non-native English Speakers across the Curriculum* (Jossey-Bass), *Language Awareness: A History and Implementations* (Amsterdam University Press), and *Revisiting the Past through Rhetorics of Memory and Amnesia* (Cambridge Scholars Press).

Darina M. Slattery
Dr. Darina M. Slattery is Head of Technical Communication and Instructional Design at the University of Limerick, Ireland, and has been teaching in higher education since 1998. Her research and teaching interests include e-learning, instructional design, technical communication, and virtual teams. She has published her research in international conference proceedings and in refereed journals, including the *IEEE Transactions on Professional Communication* and the *Journal of Technical Writing and Communication*. Previously, she was Program Director for the MA in Technical Communication and E-Learning and the Graduate Certificate in Technical Writing programs.

Dan Voss
Dan Voss is a Proposal Specialist for Lockheed Martin; provides in-service training in communication, ethics, and intercultural communication; and teaches college part-time. A fellow in the Society for Technical Communication (STC), Voss is recognized for his publications and presentations on diverse subjects at STC's international conference. With Lori Allen, he coauthored *Ethics in Technical Communication: Shades of Gray*. He coauthored four research articles for STC's peer-reviewed quarterly journal, including

"Ethical and Intercultural Challenges for Technical Communicators and Managers in a Shrinking Global Marketplace" with Madelyn Flammia. Voss and coauthor Bethany Aguad co-managed STC's student outreach and mentoring initiative in 2012–2014.

Dan Wang

Dan Wang is an Associate Professor of Organizational Theory at the School of Management, Harbin Institute of Technology, China. She earned her Ph.D. in management science and engineering in China and her Master of Business Administration from Brisbane Graduate School of Business at Queensland University of Technology, Australia. She was a visiting professor at Krannert School of Management, Purdue University, in 2007. Her research interests are in organizational theory and intercultural business collaboration. She has published papers in several peer-reviewed journals and at international conferences.

Yinqin Wang

Yinqin Wang is an Associate Professor of Intercultural Management at the School of Management, Harbin Institute of Technology, China. She received her M.S. in Engineering in 1989. She worked on the "Intercultural Factors in Technical Documentation" project at the Department of Technical Information, DaimlerChrysler AG, from 1996 to 2000. In 2000, she earned her Ph.D. from the Technical University of Berlin. Her research interests are in intercultural technical communication and cross-cultural management and she has published papers in these fields.

Han Yu

Han Yu is Associate Professor of Technical Communication at English Department, Kansas State University. Yu teaches technical communication, engineering communication, scientific communication, and business/workplace communication at both undergraduate and graduate levels. Her research focuses on visual communication, intercultural technical communication, scientific communication, and writing assessment/training. She has published multiple articles on these topics in various journals in the field. Han edited (with Gerald Savage) *Negotiating Cultural Encounters: Narrating Intercultural Engineering and Technical Communication*, published by Wiley-IEEE Press. She has also authored the book *The Other Kind of Funnies: Comics in Technical Communication* has been published by Baywood Publishing.

Index

Accreditation Board for Engineering and Technology (ABET), 11, 240, 250, 254
achievement (as in achievement vs. ascription), 222, 232–233
affective (as in neutral vs. affective), 222, 233–234
affective domain, 152–153
American Association of Colleges and Universities, 248
American Council on Education, 248
analytical style, 29, 92, 95
Android, 71–72
Appalachian State University, 181
Aristotle, 95, 99
ascription (as in achievement vs. ascription), 222, 232–233
assessment metrics, xxxv, 57–64, 249, 257
autonomous learning, 129–134, 136, 138–142, 144–146

Behaviorism, 151–152
Bildungsroman, 138
Bloom, B. S., 152–153

cage-painting, 95
categorical imperative, 100
Chinese characters (as in text and symbols), 38
Ching Yun University (CYU), 181–184, 187–189, 195
coaching, 232
cognitive domain, 152–153
cognitivism, 151, 154

collaboration, 150–151, 155–156, 162, 164–165, 168–169, 174–175, 178, 196
collectivism (as in individualism vs. collectivism), xxiii, 222
Colors, 50, 52
communication design, 48–51, 56–57, 64
communicatively effective image, 47–49, 51–53, 55–56, 58–59, 62–64
communitarianism (as in individualism vs. communitarianism), 222
community outreach, 50, 54
concentric-ring model (value analysis), 103–106, 110, 116
Confucian dynamism, 221, 223
Confucianism, 99
constructivism, 130–131, 151
content organization, 19–21, 25, 29–30, 32, 42–43. *See also* text organization
context, high- and low-, 20, 26, 29, 38–39, 42, 203
contextual relativism, 102
critical incidents, 227–230, 235
cross-cultural comparisons, xx–xxi, 24, 26–27, 29, 35–36, 40–41, 185
cross-cultural dialogs, 208, 227–229, 236
cross-cultural resonance, 59, 63
cross-cultural virtual teams (CCVTs), 4, 7, 9–10
Crystal, David, 4–5
cultural competence, xxiii, 192, 241
cultural dimensions, 24, 56, 116
cultural heuristics, xxxv, 219–220, 222–224, 226–227, 230, 232, 234, 237

Teaching and Training for Global Engineering: Perspectives on Culture and Professional Communication Practices, First Edition. Edited by Kirk St.Amant and Madelyn Flammia.
© 2016 The Institute of Electrical and Electronics Engineers, Inc. Published 2016 by John Wiley & Sons, Inc.

cultural models, 202, 208–209, 215, 252
cultural narratives, 219–220, 227, 230, 234–236
cultural self-awareness, 97, 228
cultural theories, 202, 206–208
cultural variables, 97, 201
culturally specific rhetorical strategies, 229
curricular experiences, 129–130, 133–134, 136–142, 144–146

Delphi technique, 251
Developmental Model of Intercultural Sensitivity, 252
diffuse (as in specific vs. diffuse), 222
document design, xxxii, 20–22, 53

educational stories, 138
educational trajectories, 132–133, 138, 140–141, 144–146
e-learning, xxxiv, 151–152, 155–156, 158–164, 166–169, 205
engineering students, 5, 93, 96–98, 219, 251, 255, 257–258
enterprise-level servers, 71
ethical issues, 91
ethics, xxxiii, 91–104, 106–110, 112–114, 116–121
ethnocentrism, 111
evaluative acts, 144
evangelism, 78–79
experiential learning, 174–182, 190, 195

Facebook, 14, 132, 155, 159, 182, 185, 187, 192
femininity (as in masculinity vs. femininity), 221
functionalist paradigm, 225–226

Gagne, R. M., 152–153
General Public License (GPL), 72, 84
Georgia Institute of Technology (Georgia Tech), 240, 256–257
Global Classroom Project (GCP), 175, 178–179
global competence, 5, 240, 246, 252, 254, 258
global virtual team teaching module, 180, 190
global virtual teams, xxii, xxxiv, 164, 173–176, 178–180, 190, 194–195

globalization, xxxii, 5, 47–48, 82, 92–94, 130, 155–156, 158, 195, 202, 204–206, 216
glocal, xxxiii, 70, 81–84, 88

Hall, Edward T., 203, 206, 209, 220–221, 223, 225–226, 230
Hampden-Turner, Charles, 220, 222
handheld devices, 70–73, 77
hierarchical model (value analysis), 103, 105–106, 110–112, 116
high-context, 20, 26, 29, 34, 38–39, 42, 203
Hofstede, Geert, 203, 205–206, 209

Icograda's Design Education Manifesto, 64
image evaluation, 53, 56–59
individualism (as in individualism vs. collectivism; also as in individualism vs. communitarianism), 221–223
information and communication technologies (ICTs), 127–129, 131–132, 139–142, 146
information design, 150–151, 156, 162, 163, 168
Institute of International Education, 242
instructional design, 150–151, 153–154, 156–158, 160–163, 166–169
intercultural assessment research, xxxv–xxxvi, 202, 206–212, 214, 244–245
intercultural communication, xxx, xxxv, 22, 92, 127, 178, 202–203, 219–220, 230, 234–236
intercultural competence, 150, 192, 241, 244, 246–247, 249–254, 258
intercultural learning, xxxv, 98, 141, 145, 228, 239, 241–248, 250, 254–259
intercultural maturity model, 252
intercultural technical communication, 24, 42, 92–93, 114, 119–121, 192, 223, 228–229
International Council of Design (formerly International Council of Graphic Design Associations), 64
international technical communication, xxxi, 19, 192, 201–202, 204–206, 214–215, 228
internationalization, 160, 162–163, 202, 204–207, 215

interpretive paradigm, 225–226, 230–231
Irish information technology industry, 158–160

Joint Task Force on Student Learning report, 176, 178

Kalman, Tibor, 50
Kana, 38
Kanji, 38
Kant, Immanuel, 99–100, 103
knowledge construction, 130, 142
Kohl, John, 157
Kolb, David A., 176

language awareness, 5, 8, 11, 13, 15
learning pedagogies, 150–156, 164
Lewin's Experiential Learning Model, 176–177, 182, 185
linguistic pragmatics, 14
linguistic relativity, 5–6, 21
linguistics, xxxii, 3–15
Linux, xxxiii, 69–88
Linux-based operating system, 69–88
localization, 53, 150, 158–160, 162–163, 169, 202–207, 210, 215
Localization Industry Standards Association (LISA), 204, 206
the logic model framework, 241–243
long-term orientation (as in long-term vs. short-term orientation), 221
low-context, 20, 26, 29, 34, 38–39, 42, 203
Lumina Foundation, 240

masculinity (as in masculinity vs. femininity), 221
Mercer University's Master of Science in Technical Communication Management (MSTCO), 201–202, 204–205, 207, 215
Microsoft, 69–72, 76–79, 83
Mill, John Stuart, 99, 103, 106
modeling, 154
module-based approach, 180
monolingualism, 5, 7–8
Multiliteracies for a Digital Age, 78

National Academy of Engineering, 240
neutral (as in neutral vs. affective), 222, 233–234

Nicomachean Ethics, 99
non-formal learning, 255
North Dakota State University, 9

open source community practices, xxxiii, 47, 49, 69, 73
open stories, 227, 229–230
Oracle VM Virtual Box, 84–87
outcomes assessment, 239–241, 244, 246, 248, 252

particularism (as in universalism vs. particularism), 222
pictographic language, 34, 38, 43
power distance, 221
problem-based collaborative learning, 136, 139
problem-based learning (PBL), 131, 134–136, 141, 143–144
process flow for value analysis, 92, 114–116
professional communicators, 156–158, 160, 163, 201–202
project learning applications (PLA), 131, 136, 144
psychomotor domain, 152–153
Purdue University, 257
Python (computer language), 76

Rand, Paul, 49
redundant relationship, 223
relational style, 99
relativism, 92, 94–95, 98–102, 113, 115
rhetorical awareness, 150–151, 156–157
Ricoeur, Paul, 138

scaffolding, 154
Selber, Stuart, 78, 83–84
server markets, 71
short-term orientation (as in long-term vs. short-term orientation), 221
Skinner, B. F., 151–152
Skype, 14, 182, 187
SMART objectives, 247
socially conscious communication design, 47, 49, 51, 56
sociolinguistics, 14
specific (as in specific vs. diffuse), 222
stereotyping, 111
sustainability, 59–60, 63

target language(s), 19–20
technical communication, xvii, xxxiii–xxxiv, 22, 24–25, 30, 42, 69, 80, 83–84, 86–88, 92, 103, 106–107, 114, 117, 119121, 149–150, 156, 158–163, 167, 169
text organization, 23. *See also* content organization
text-graphic relationships, 33–35, 37–38, 41–42
thought pattern, 19–21, 23, 29–30, 32, 34, 38–39, 41–43
Trans-Atlantic & Pacific Project (TAPP), 4, 10, 13, 15–16
transdisciplinary approach, 219, 226–227, 230, 234, 237
translation, 3, 8–15, 20, 157–158, 180, 182, 189, 193, 196, 202, 204, 206–207, 210–211, 213–215
Trompenaars, Fons, 206, 220, 222, 230, 232–233
Tuning Project, 240

Ubuntu, 74–75, 82, 84, 87
uncertainty avoidance, 221, 232

universalism (as in universalism vs. particularism), 222
universalism (as in universalism vs. relativism), 92, 94–95, 100–101
Universidad de Extremadura, 80–81, 85
University of Guadalajara, xxxiv, 127, 142
University of Limerick MA Program, xxxiv, 158, 160–161, 164–165, 169
Utilitarianism, 99

value analysis, 92, 95, 98, 102–106, 109, 111, 113–119
verbal credibility, 52, 63
virtual teams, 156, 164–165
virtue ethics, 95, 99
visual communication, 21, 33–34, 37, 40, 205, 207
visual credibility, 52, 62
von Humboldt, Wilhelm, 21

Yerevan State Linguistic University (YSLU), 183–184, 186, 188
Yi Jing (The Book of Changes), 22, 38

Books in the
IEEE PCS PROFESSIONAL ENGINEERING COMMUNICATION SERIES

Sponsored by IEEE Professional Communication Society

Series Editor: Traci Nathans-Kelly

This series from IEEE's Professional Communication Society addresses professional communication elements, techniques, concerns, and issues. Created for engineers, technicians, academic administration/faculty, students, and technical communicators in related industries, this series meets a need for a targeted set of materials that focus on very real, daily, on-site communication needs. Using examples and expertise gleaned from engineers and their colleagues, this series aims to produce practical resources for today's professionals and pre-professionals.

Information Overload: An International Challenge for Professional Engineers and Technical Communicators · Judith B. Strother, Jan M. Ulijn, and Zohra Fazal

Negotiating Cultural Encounters: Narrating Intercultural Engineering and Technical Communication · Han Yu and Gerald Savage

Slide Rules: Design, Build, and Archive Presentations in the Engineering and Technical Fields · Traci Nathans-Kelly and Christine G. Nicometo

A Scientific Approach to Writing for Engineers & Scientists · Robert E. Berger

Engineer Your Own Success: 7 Key Elements to Creating an Extraordinary Engineering Career · Anthony Fasano

International Virtual Teams: Engineering Global Success · Pam Estes Brewer

Communication Practices in Engineering, Manufacturing, and Research for Food and Water Safety · David Wright

Teaching and Training For Global Engineering: Perspectives on Culture and Professional Communication Practices · Kirk St.Amant and Madelyn Flammia